Josefer Ranchel da Silva Neves
Rosivaldo C. Gama Júnior
Gerlany F. S. Pereira

Experimentation in physics teaching

Josefer Ranchel da Silva Neves
Rosivaldo C. Gama Júnior
Gerlany F. S. Pereira

Experimentation in physics teaching

Student experiences in the 9th year of primary school

ScienciaScripts

Imprint

Cover image: www.ingimage.com

This book is a translation from the original published under ISBN 978-620-2-03188-2.

Publisher:
Sciencia Scripts
is a trademark of
Dodo Books Indian Ocean Ltd. and OmniScriptum S.R.L publishing group

120 High Road, East Finchley, London, N2 9ED, United Kingdom
Str. Armeneasca 28/1, office 1, Chisinau MD-2012, Republic of Moldova, Europe
Managing Directors: Ieva Konstantinova, Victoria Ursu
info@omniscriptum.com

Printed at: see last page
ISBN: 978-620-6-76186-0

DEDICATORY

To my father Francisco Neves da Silva and my mothers Antônia da Silva Neves and Osmandina da Silva Neves and Orlandina da Silva Neves.
(JOSEFER RANCHEL DA SILVA NEVES)

My daughters Lívia Feitosa Furtado Gama and Eloísa Feitosa Furtado Gama and my wife Cherliane Feitosa Furtado.
(ROSIVALDO CARVALHO GAMA JÚNIOR)

To the science teachers who seek to innovate their methodological practices in the midst of a country contaminated by corruption, but who still believe (as I do) that education can change this situation! **(GERLANY DE FÁTIMA DOS SANTOS PEREIRA)**

ACKNOWLEDGEMENTS

I would like to thank my family members who directly or indirectly helped me to complete my degree.

I would like to thank all the teachers who have helped me acquire knowledge during my training, especially my supervisor, who has always encouraged me to look for the best ways to study physics, and my co-supervisor, who has always been willing to contribute. In other words, I would like to thank both supervisors, who have made a significant contribution to this work. I learnt a lot from you.

I would like to thank the students who took part in the research, the class teacher who was always helpful and made herself available to contribute to the research, and the school staff who welcomed us warmly. (**JOSEFER RANCHEL DA SILVA NEVES**)

I thank God for giving me moments of dedication to build this work.

Thanks to all the people who contributed to the publication of this book. (**ROSIVALDO CARVALHO GAMA JÚNIOR**)

Thank you to everyone who made it possible to publish this book! In particular, to Rafael Viana, our editor, and to NEA, which made this possible. (**GERLANY DE FÁTIMA DOS SANTOS PEREIRA**)

The only place where success comes before work is in the dictionary.

(Albert Einstein)

SUMMARY

The aim of this book was to analyse the effects of experimentation on students' level of interest in the process of teaching and learning physics in the 9th year of primary school. The research was carried out in a state school in the municipality of Santana-Amapá, with four 9th grade classes. The methodological approach used was action research. The data was collected through participant observation in physics classes and the application of semi-structured questionnaires with the students. The data was analysed using a qualitative-quantitative approach with the construction of categories of analysis and the creation of graphs and charts to represent the data found. The results show that scientific experimentation associated with physics lessons is a fruitful methodology for tackling the challenges teachers face in providing meaningful learning for students. On the other hand, teachers must bear in mind that their work in the classroom will be much greater if they try to overcome the mere "transmission of information" that predominates in natural sciences classes.

Key words: Teaching and Learning. Science Teaching. Teaching Practice.

SUMMARY

CHAPTER 1	6
CHAPTER 2	8
CHAPTER 3	19
CHAPTER 4	26
CHAPTER 5	34
CHAPTER 6	66

CHAPTER 1

INTRODUCTION

It can be seen that teaching activities in the final grades of primary school in public schools in the municipality of Santana-AP, related to the teaching of natural sciences, are generally not practical, with predominantly theoretical lessons. This has a direct impact on the level of student (lack of) interest in introductory physics teaching at this stage of schooling.

In this sense, throughout their school life, students acquire a certain aversion to physical concepts, especially when they enter secondary school. This is why the following question is necessary: to what extent do physics lessons taught in a theoretical-practical way influence the level of interest of students in the 9th year of the final years of primary school?

When teaching physics to 9th graders, the teacher needs to have teaching materials that make the scientific content more attractive to the student, with the aim of adding to and/or stimulating the student's taste for science in later high school years.

In this way, it is assumed that by associating theoretical lessons with experimentation, the teacher will develop learning methodologies in the classroom that will reflect on the level of student interest in the teaching and learning process in Physics.

The National Curriculum Parameters (PCN) (BRASIL, 1998) point out that, due to the scope and nature of the objects of study in the sciences, there are notable possibilities for developing the subject in a dynamic way, guiding the teacher's work towards the development of methodologies capable of making the content meaningful to the student's reality.

Furman (2009) elucidates that when teachers seek teaching methodologies aimed at making concepts more attractive to students' learning, such as associating theoretical lessons with experimentation that is appropriate to the students' reality. As well as making it more attractive for them to learn the conceptual part of science, it also favours the development of procedural skills that are part of the very nature of scientific knowledge: such as observation, recording variables, reflection and theoretical proof of the content taught (which can be tested through practical activities in the classroom) (FURMAN, 2009).

In this context, research into the teaching of science has emphasised that there is great concern to ensure that the teaching of this subject is effective, so that the teacher works on valid and correct concepts, following a methodology that is appropriate to the student's level of cognitive development, with the aim of contributing to their learning (NARDI, 2001).

With this in mind, the main objective of this book is to analyse the impact of the theoretical-practical nature on the level of interest of students in the teaching and learning process of physics in the 9th year of primary school in a public school in the municipality of Santana-AP.

This objective unfolds into the following: to verify the methodological resources used in the teaching of physics classes; to diagnose the level of student interest in introductory physics teaching when the content is taught without a theoretical-practical relationship; to identify the reflexes in relation to the level of student interest in introductory physics teaching when the content is taught in a theoretical-practical way and to suggest a theoretical-practical class proposing the importance of experimentation in relation to the level of student interest in physics.

CHAPTER 2

THE CHALLENGES ENCOUNTERED BY TEACHERS WHEN TRYING TO USE INNOVATIVE METHODS IN PHYSICS TEACHING

Education has undergone various transformations throughout history, because depending on the context in which it is observed, it can be seen that individuals are trained to meet the cultural, political and economic aspirations of a society at a given time. In this context, in the current social conjuncture, theorists emphasise that:

> In recent decades, as a result of the universalisation of access to basic schooling and changes in society and the world of work, schools have been called upon to modify their content, objectives and teaching methodologies, as well as to promote the socialisation of students in the cultural context of which they are a part, to qualify their practices and their effectiveness as an instrument for the moral and intellectual development of students (PORTO; RAMOS; GOULART, 2009. p. 6).

It is up to schools and teachers to improve their competences and skills so that students not only assimilate the contents of the curriculum. But that they develop a critical sense based on effective, contextualised teaching that can actively intervene in social relations, preparing students to exercise their citizenship.

To this end, developing more attractive and contextualised learning methodologies for 9th grade students to critically construct their knowledge of physics requires teachers to overcome some challenges in relation to the way in which the teaching of this science has predominantly been presented over the years:

> Teaching that is primarily geared towards preparing for university entrance exams, supported by the indiscriminate use of textbooks or similar materials and an excessive emphasis on solving purely memorised and algebraic exercises. Teaching that presents physics as a compartmentalised, segmented, ready-made, finished, unchangeable science (MEGID NETO; PACHECO, 2001. p. 17).

This makes it more "comfortable" for teachers to use textbooks and solve exercises in class without adequate contextualisation and to ask students to "memorise" scientific laws and theories, which "[...] should also be considered that many educators learned science in this way and unconsciously teach in a similar way " (SOUZA et al., 2012, p. 2).

On the other hand, the lack of laboratories in most schools that allow teachers to innovate their lessons, such as planning an experiment that is appropriate to the student's reality and level of cognitive development, ends up making it difficult to do so, "[...] where there are several records that in Brazil science teaching in general, and physics teaching in particular, has not been enjoyable, curiosity-provoking or creative, a condition that affects interest and learning in physics" (RODRIGUES; TEIXEIRA, 2011, p. 2).

It is worth emphasising that, despite the above questions and other factors that present challenges for science teachers to innovate their lessons, such as the reality of the school and the curriculum, which are elements that contribute to this (SOUZA et al., 2012.), it should only be pointed out that it is essential for 9th grade students to build a good foundation in relation to physical concepts, in order to prepare them to enter high school.

In this way, seeking methodologies that encourage students' interest in physical concepts requires teachers to overcome these and other challenges that influence their teaching practice in public schools, where "[...] using teaching strategies that allow children to come into contact with material objects and develop experiments contributes to learning physics in a playful and enjoyable way" (SOUZA et al., 2012, p. 4).

2.1 THE NEED TO ADAPT CONTENT AND METHOD IN SCIENCE LESSONS

Faced with the constant technological and scientific transformations that are taking place and moulding contemporary society, "[...] creating autonomous students who know how to think, make their own decisions and study on their own is one of the goals of teaching" (CARVALHO et al., 2007, p. 29). This becomes relevant due to the constant scientific and technological advances of today.

In this way, students are increasingly required to develop competences and skills in order to facilitate their performance and guide their actions consciously in this world of transformations. "[...] Today the world reveals the need for man to be creative, interpretative and with agile reasoning, reflecting on his survival, work, study and where he lives" (MEDEIROS, 2010, p. 165).

These objectives require the constant efforts of parents, school staff and, above all, teachers in the classroom when training students, as it is important to bear in mind the type of student you want to train. Delizoicov; Angotti and Pernambuco (2007, p.12) emphasise that, "The challenge is to educate children and young people, providing them with human, cultural, scientific and technological development, so that they acquire the conditions to face the demands of the contemporary world". In this sense, the National Curriculum Parameters (PCN) propose that:

> [...] showing science as knowledge that contributes to understanding the world and its transformations, to recognising man as part of the universe and as an individual, is the goal that is proposed for teaching the area in primary school (BRASIL, 1998, p. 21).

Science teaching in primary school is concerned with preparing students to understand scientific phenomena and their applications in everyday life, or their direct relationship with the body, the environment and the transformations produced in society associated with technology, broadening their present possibility of social participation.

However, we are faced with low levels of school performance, many students dropping out and dissatisfaction with the teaching methodology used in most formal public schools (OLIVEIRA et al., 2010).

The indicators of the Programme for International Student Assessment (PISA) (ÉPOCA, 2013) show the low level of scientific and technological knowledge of Brazilian students among the countries participating in the survey. In this sense, it is clear that didactic strategies involving experimentation can be relevant mechanisms for introducing students to a more lively and dynamic science.

In this way, we can see the adversities that science teachers will have to overcome in the course of their lessons, since teaching a curricular component as complex as physics, with scarce teaching materials and methods, as well as the lack of adequate contextualisation of the content to the students' reality, would make it difficult to build students' autonomy and critical spirit, and could have a direct impact on the levels of approximation between students and the subject of physics in the 9th grade. According to Oliveira et al. (2010, p. 31):

> The problem lies in the concept of school teaching centred on passing on knowledge from one person to another, as if it were a substance. Consequently, the role of the teacher would need to be reformulated, and training in the use of experimentation in

the context of research-based education could be part of this reformulation.

According to Carvalho et al. (2007, p. 20) "The problem is the driving force behind the students' varied actions: it motivates, challenges, arouses interest and generates discussion". This assumption can be explored by associating a theoretical lesson with a practical activity.

The teacher will create an environment where the students can give their opinions and make inferences about the theory being worked on and, after this stage, check in the course of the experiment whether their statements are consistent with the laws and theories of science, in an attempt to move beyond teaching that is predominantly centred on passing on information.

In view of this, the school needs to review its role in the formation of citizens, checking whether its teaching plans and teaching staff are working in favour of school success. In other words, whether the educational institution fulfils its social role effectively in accordance with the students' reality and its own structural reality:

> Critics of contemporary education point out that our educational system practises a reproductive learning model based on copying, and that today's school functions almost identically to the school of decades past, with teachers gradually transferring knowledge to students (OLIVEIRA et al., 2010, p. 31).

This need to adapt the content stems from the fact that students bring previous knowledge to school, built up from their social interactions, according to Ausubel (2003). And also because of the wealth of information available to citizens in the most diverse communication and information vehicles, giving them access to content even before it is worked on at school.

In this way, conceiving of the student as a subject who brings with them knowledge from their experience clashes with the traditional stance, "[...] this view of teaching and learning had at its core a variety of epistemological stances, for example: the student being a tabula rasa, that is, not possessing previous ideas when they arrive at school" (ATAIDE; SILVA, 2011, p. 173).

The National Curriculum Parameters (BRASIL, 1998) emphasise that students outside of school develop explanations for natural phenomena and technological products, where these explanations satisfy their immediate questions and curiosity. However, in school, when they come into direct contact with scientific concepts, their explanations often require a re-signification (BRASIL, 1998).

Using problem situations, the teacher exposes the content in such a way as to challenge the students to answer questions about the topic being worked on, making the most of their previous knowledge and allowing for dialogue during the course of the lessons, thus favouring a change in the way the content is approached in the classroom. Thus:

> Whenever possible, teachers need to be innovators, but when they try to change the way they work and assess, they must be aware that the main objective of this innovation is the quality of teaching, and so they must never forget their serious commitment to learning (SOUZA; HEINECK, 2006, p. 7).

However, when teachers recognise that their role in the classroom goes beyond the mere "transmission" of content, they set out to face the adversities of the school environment in order to innovate their lessons, in which "[...] the teacher is no longer seen as an 'applicator' of pre-established techniques and/or models" (MARTINS, 2009, p. 2), but as a builder of knowledge, re-signifying students' previous knowledge and enabling scientifically-based constructions. In these terms:

> A constructivist teacher is expected to do much more than just know how to explain the subject and have a good relationship with the students. Their lessons need to be creative and provide space for the learning situations necessary for students to construct their knowledge (CARVALHO et al., 2007, p. 35).

In the case of science teachers in the 9th year of primary school, by emphasising a teaching methodology combined with experimentation, they will be helping students to acquire a taste for physics. At the same time, they will be able to approach this curricular component in later years of secondary school in a more flexible and satisfactory way.

2.2 EXPERIMENTATION ASSOCIATED WITH TEACHING PRACTICE IN PHYSICS: TEACHING METHODOLOGICAL AND PROCEDURAL RELEVANCE FOR STUDENTS

In today's information and communication society, science teaching has increasingly demanded more from teachers. Conceptual lessons are no longer holding the attention of students who today enjoy the advanced audiovisual devices characteristic of modernity, such as mobile phones, notebooks, games and other technological tools that make their lives more dynamic.

In this sense, it is known that Physics-related content offers a range of experiments that can be explored in the classroom in order to awaken students' taste for Science. Both in the space of laboratories and the use of alternative materials, thus requiring a greater deployment of the teacher when planning their teaching actions in order to develop differentiated activities in the teaching of Physics.

Thus, "[...] Physics as a subject in the Brazilian school curriculum was introduced in 1837, with the foundation of the Pedro II school in Rio de Janeiro" (MEGID NETO; PACHECO, 2001, p. 17). However, over the almost 180 years since this subject was introduced into the teaching and learning process in Brazilian schools, the teaching characteristics of this science have often lacked experimentation that is appropriate to the student's reality.

Experimentation in Physics stems from the student's interaction with a given piece of knowledge that becomes "tangible" when worked on in experimental situations. In other words, methodological actions carried out in the classroom, in laboratories, in the natural environment and in other contexts that allow students to see in practice the concepts and laws that govern the sciences:

> We realise that experimental activities have become necessary during the teaching-learning process, as they can be another pedagogical tool to motivate and instigate students, making them active agents in the construction of their knowledge (SOUZA et al., 2012, p. 4).

In this way, the didactic potential of a theoretical lesson associated with experimentation depends very much on the teacher's sensitivity and creativity when it comes to looking for practical activities that attract the students' attention and provide more comprehensible learning situations aimed at the students' long-term education, making learning physics more attractive and enjoyable:

> Public elementary schools give greater importance to phenomena related to physics only in the 9th year of primary schools, because the subject where it should be included (science) is more focused on the biological sciences (SOUZA et al., 2012, p. 1).

"Since physics is an inevitably quantitative science, it consequently depends on measurement procedures" (LABURÚ; MOURA; SALES, 2010. p. 2). Because theoretical concepts and physical quantities form the building blocks of this science.

Hence the importance of choosing the type of methodology or teaching strategy that the teacher intends to develop in the classroom so that students have this first contact with physics and know how to deal with the problems involving scientific phenomena. Bearing in mind that 9th grade is the time when students will have their first direct interaction with physical laws and phenomena as a "shared" and introductory subject.

It is worth emphasising that lessons associated with experiments are not the only alternatives to be applied by the teacher in the classroom:

> Experiments and observations allow us to build up some ideas about phenomena, but they leave numerous gaps that need to be filled and deepened with information that experiments alone are unable to provide, but which can be provided by a teacher, a text or an expert (FURMAN, 2009, p. 17).

It is therefore up to the teacher to look for other resources that can add to the student's education and arouse interest in science classes, such as books, scientific journals, specialised science sites on the Internet, libraries, games, educational software and other teaching materials that contribute to practical classes.

From a procedural point of view, experimentation in physics, as well as being a didactic tool that subsidises theoretical lessons, can also stimulate procedural skills in students:

> Identifying the competences we want to teach when we carry out an activity with our students is fundamental if we want practical lessons to stop being simply a time to get "hands-on".
>
> become opportunities to put minds into action (FURMAN, 2009, p. 14).

Based on the idea that the teacher should encourage and/or instigate the students' ability to observe, record, reflect and construct knowledge based on real situations:

> The latter is associated with doing science, that is, with the actions that support the investigative activity of experimental work conducted by planning in which the data is analysed and used to support certain conclusions (LABURÚ; MOURA; SALES, 2010, p. 2).

On the other hand, according to Souza et al. (2012, p.2) in relation to theoretical-practical classes, "[...] it is believed to be a way of promoting scientific literacy, as it puts students in touch with science and can awaken their interest in it". Therefore, at all times when learning Physics through experimental situations, in order to make lessons more attractive to 9th graders, the teacher's presence is essential, both in conducting the experiment and in theoretically helping the students to build procedural competences. These stages of experimentation are also characteristics of the very nature of Science, which in such a way exploit the student's creativity and greater involvement in the lesson, and can reflect on their levels of interest in the discipline in question.

2.2.1 CONTRIBUTIONS TO THE STUDENT'S EDUCATION WHEN HANDLING EXPERIMENTS

Among the activities that teachers can work on in the classroom with their students are practical lessons. Most of the time, these take place in laboratories with the aid of technological equipment that gives students the opportunity to visualise phenomena and relate them through observations to the theories being studied. In this sense, Imbernon et al. (2009, p. 84) give us a classification of teaching methodology related to experimentation:

> Experimentation activities can be closed demonstrations or open demonstrations/observations. While closed demonstrations are mainly characterised by the simple illustration of a given phenomenon, they are a teacher-centred activity. Open demonstration/observation activities incorporate other elements, presenting

> greater openness and flexibility for discussions that can allow for a deeper understanding of the conceptual and practical aspects related to the equipment [...].

In this way, the so-called closed demonstration activities do little to problematise the student, where the teacher presents the activity in an illustrative way, "[...] according to the traditional school, experience appeared only after the explanation of knowledge in a theoretical way, in order to memorise and prove the information given" (ZOMPERO; PASSOS; CARVALHO, 2012, p. 45).

On the other hand, by making the experimentation teacher-centred, the students don't feel active or able to cooperate with the construction of the concepts being worked on because the activity is "ready-made" and they just have to carry it out:

> When the student is given a "script" for a laboratory lesson (or field research), where the experiments, results and conclusions are like a "cake recipe" whose aim is to answer what the teacher expects, there is no room for a "new look" or different answers; in this situation, prior knowledge is disregarded (TAUCEDA; NUNES; DEL PINO, 2011, p. 137-138).

Thus, the idea that teachers need to constantly analyse and rethink their actions in order to see if the proposed activities allow students to reconstruct knowledge is defended, because "[...] this methodology reflects a simplistic view of the possibility of 'doing science' through laboratory classes at school" (TAUCEDA; NUNES; DEL PINO, 2011, p. 138). In this sense, the teacher ceases to be the "holder" or the individual responsible for the assimilation of knowledge by the students and becomes a guide committed to a path, building the pedagogical work collectively. In these terms:

> Science lessons can and should be planned so that students go beyond contemplative action and move towards reflection and the search for explanations, because this is how students will have the chance to relate objects and events and express their ideas (CARVALHO et al., 2007, p. 21).

However, it should be emphasised that if the teacher's lesson aims to revise the concepts covered or to encourage students to expand on their previous concepts, he or she can work with closed demonstration activities, but always with the aim of making a significant contribution to students' learning. When open-ended demonstration activities are proposed for teaching, students begin to develop competences that are relevant to their education. These competences can be acquired through the manipulation of experiments:

> As a source of research into phenomena and their transformations, the experiment becomes more important the more the students participate in making their guide or protocol, carry out the actions on the materials themselves and discuss the results, prepare how to organise the notes and carry them out (BRASIL, 1998, p. 80).

In this sense, the procedural competences that students are encouraged to develop, discussed in the previous topic, such as the ability to observe, record variables and reflect on phenomena, will be considerably stimulated by open-ended activities. These activities go beyond the students' mere contemplation of the phenomena, allowing them to carry them out and gain an idea of all the practical processes involved. Thus:

> When students carry out experiments and understand or try to understand the knowledge presented, seeking to explain the phenomena that surround them, the theoretical content has a strength that makes it indelible, as opposed to the memorised content which, unfortunately, disappears without a trace (IMBERNON et al., 2009. p. 86).

To this end, Lima et al. (2009, p. 8) reinforce the following: "In addition, in the manipulation of material reality, young people are confronted with aspects of knowledge Physics that can hardly be understood in their complexity simply by exposure to theoretical descriptions". Faced with this confrontation, the student observes and tries to solve the problems proposed.

The teacher, faced with these situations, subsidises the students to make the appropriate notes, questioning them to give their opinion and manipulate the reality confronted by the experiments. In these situations "[...] the playful, enlightening and challenging nature of experimentation favours the development of students' self-esteem, their appreciation of the educational process in which they are involved, as well as their ability to operate cooperatively" (LIMA et al., 2009, p. 8).The so-called open demonstration activities allow the teacher and students to raise hypotheses that can be answered within the experimental activities carried out by the students:

> It is the teacher who proposes problems to be solved, which will generate ideas that, when discussed, will allow previous knowledge to be expanded, promotes the

> opportunity for reflection, going beyond purely practical activities; establishes collaborative working methods and a classroom environment in which all ideas are respected (CARVALHO et al., 2007, p. 36).

In this process of discussion, students can interact, dialogue and give their opinion at the moment of learning, when their inferences are valued and, being the holder of all the processes that involve the experimental activity, their autonomy and criticality are stimulated, but "[...] for this, the teacher must propose a problem to the students, a script with objectives, the procedures, notes of the data and, finally, a conclusion" (ZOMPERO; PASSOS; CARVALHO, 2012, p. 46).

Therefore, when working with experimentation as a methodology in the process of teaching and learning physics in the 9th year of primary school, teachers need to analyse whether the practical activity allows students to use all the technical procedures involved in carrying it out. Always taking care to adapt the teaching method to the grade and cognitive development that will be involved in this type of activity, where the teacher in the contemporary context is the guide who will lead the students to develop their competences and skills.

CHAPTER 3

METHODOLOGICAL PATHS

3.1 AREA OF WORK AND NATURE OF RESEARCH

The research was carried out at the Joanira Del Castillo State School, located at Avenida Pedro Teixeira, 194, in the Nova Brasília neighbourhood in the municipality of Santana, state of Amapá, Brazil. This school was set up by SEED decree 319/80 on 22 July 1980, with the former Federal Territory of Amapá as its sponsor at the time, with the aim of catering for students in grades 1 to 4 of primary school.

It began operating on a temporary basis, in the former Independence Sports Club, from 1977 to 1979, under the name Annex II Barroso Tostes. Around 1983, there was a need to open places for grades 5 to 8 due to high demand from the community, as the existing schools at the time were insufficient to meet demand.

Before its creation in 1980, the area where the school now stands was completely uninhabited (the area that today comprises the Nova Brasília neighbourhood was called Vila Maia). Initially, the school received a minimum number of classrooms to ensure its immediate operation. In the mid-1990s, there was a small extension to the building in which the administrative block was built with ten classrooms.

From a structural point of view, the school currently has a library, a secretariat, a video room, a room for teachers to plan their lessons, a room for special students, a room for school supervision and guidance, a room for the school cashier, a room for the school principal, a kitchen, toilets for students and other staff.

The most recent achievement was the implementation of a computer lab for the students. Several projects are already traditional at the school, such as the cultural project "Cara Nossa" (Our Face), the musical project "Meu Primeiro Acorde" (My First Chord), and the project aimed at students in Youth and Adult Education (EJA) entitled "A sétima arte (cinema at school)" (The seventh art).

Today, the school carries out its pedagogical activities both during the day and at night. Fourteen classrooms are open in the morning, one of which caters for special education and the others for 6th and 7th grade classes. In the afternoon there are also fourteen classrooms catering for 8th and 9th graders. In the evening, the school has five classes catering for EJA students.

In this sense, this study, carried out at the aforementioned school, was configured as action research. With regard to action research, Barbier (2007, p. 53) emphasises that it is research aimed at "[...] serving as an instrument for social change [...] action research postulates that the production of knowledge cannot be dissociated from the efforts made to bring about change".

"If someone chooses to work with action research, they certainly have the conviction that research and action can and should go hand in hand when the aim is to transform practice" (FRANCO, 2005, p. 485). From the same perspective, Bourscheidt (2007, p. 1) emphasises that in action research "[...] both the researcher and the group being researched interact in a participatory way, developing the ideas proposed in the research plan".

In this context, action research is research that, in addition to understanding, aims to intervene in the situation in order to change it. The knowledge sought is linked to an intentional goal of changing the situation being researched. Thus, at the same time as diagnosing and analysing a given situation, action research proposes changes to the group of subjects involved that lead to an improvement in the practices analysed.

3.2 DATA COLLECTION

The research began on 1 September 2014, when the researcher presented himself at the school to the pedagogical coordinator with the document requesting the start of the activity. After this stage, the four 9th grade classes that took part in the research were randomly selected. They were labelled X, Y, Z and W.

Firstly, participant observation (GERHARDT et al., 2009) took place over a two-week period, from 1 to 12 September 2014, specifically on Mondays, Tuesdays and Fridays at all times in the four classes selected. In this way, each class had three science lessons (introduction to physics) each week.

In the second phase, on 15/09/14, 16/09/14 and 19/09/14, theoretical and practical classes were held with the students. To implement the teaching methodology in question, two timetables were used to work on it in each class. The lesson consisted of a theoretical exposition of the content (gravitation and Newton's laws) using slides and illustrations of situations from the students' day-to-day lives with images. The practical aspect of the lesson was the execution of three experiments manipulated by the students. These were: 1st) The Rope Challenge; 2nd) Trombada and 3rd) Mini-catapult, as shown in the images below.

In the first experiment, as shown in figure 1, the idea was to prove that you can't cancel out the weight force, a vertical force, by applying horizontal forces. A thick book is attached to the end of the smaller rope (E), then the other end is attached to the middle of the larger rope (H). To see if the rope is taut or not, a string (V) is held near the ends of the rope (H). The students are then challenged to stretch the rope in such a way that it touches the string.

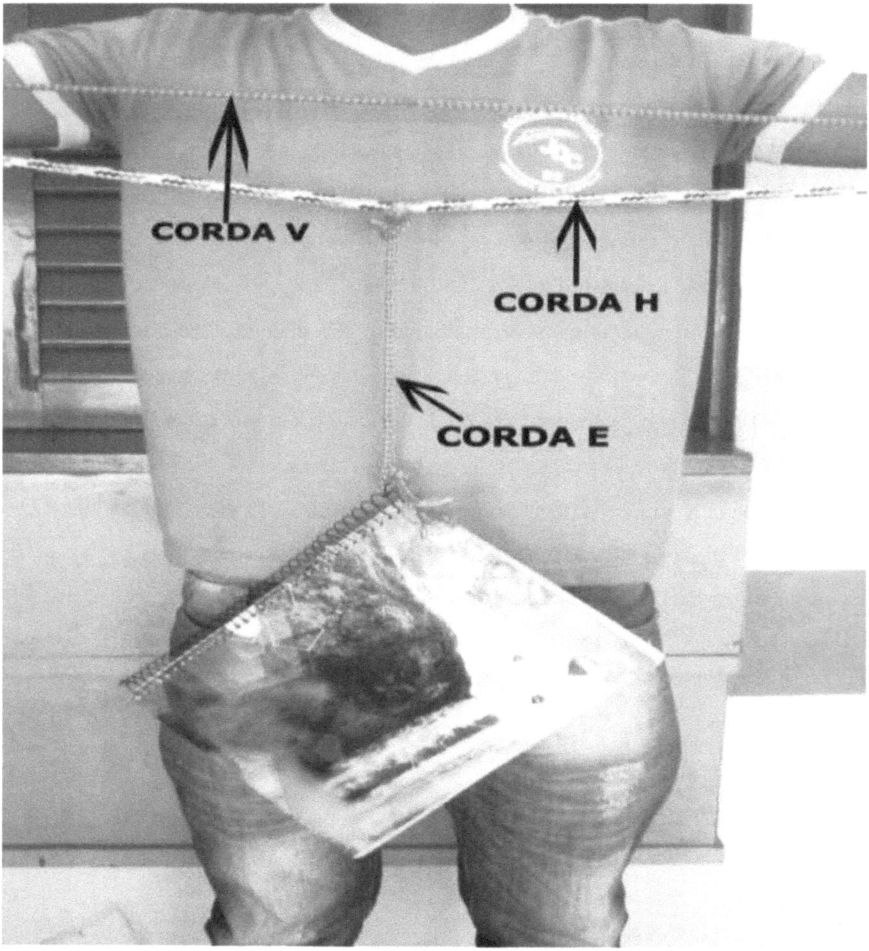

Figure 1 - Experiment 1 - **"The rope challenge".**

In the second experiment, see Figure 2, we worked on the principle of inertia, or Newton's first law, which states that an object always tends to maintain its state of motion, which can also be that of rest if there is no external force acting on it (GASPAR, 2010). The students were asked to disregard a trolley with a small ball attached to it, roll it down a ramp and bump into an obstacle, so the trolley would travel down the ramp until it hit the obstacle. When it hits the obstacle, the trolley stops; the steel ball, however, being only lightly attached to the trolley under the surface of the modelling clay, tends to continue its movement, being thrown forward.

Figure 2 - Experiment 2 "Trombada".

In the third experiment, illustrated in figure 3, a mini catapult was built in order to verify Newton's second law, which states that the acceleration acquired by a body is directly proportional to an external force applied to it and inversely proportional to its mass (GASPAR, 2010).

The students threw a matchstick and a dice with the same amount of force and then had to observe what different heights the two objects acquired, as well as the causes of this phenomenon.

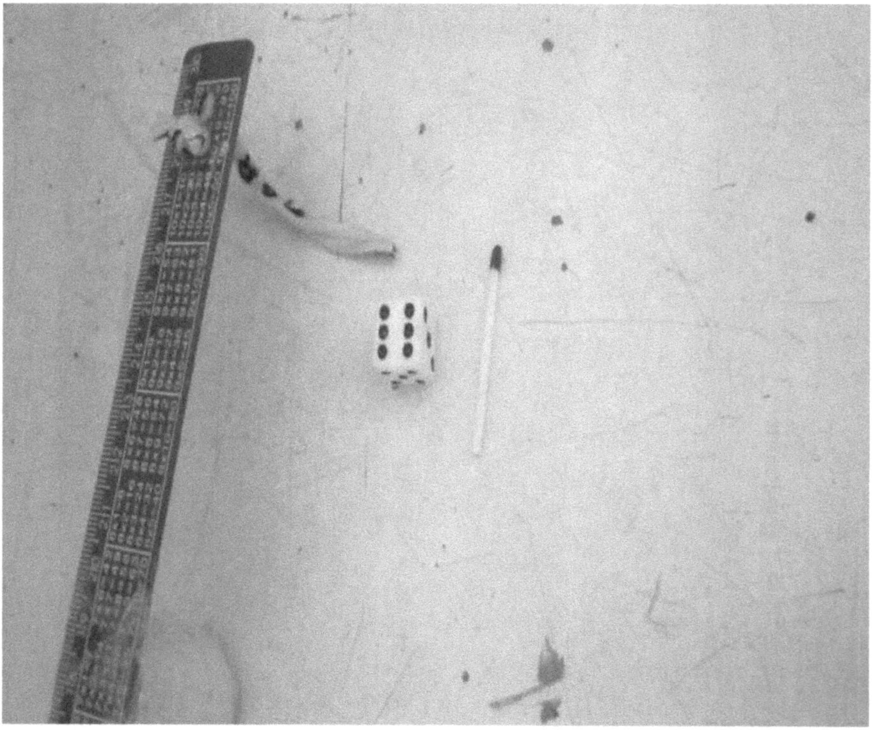

Figure 3 - Experiment 3 "Mini-catapult".

After the theoretical-practical lesson, the students were asked to solve the three-question activity with a mathematical approach to the content covered (APPENDIX1).

It should be noted that during this period of data collection, from 01/09/14 to 19/09/14, the researcher used a logbook based on the precepts of Gerhardt et al. (2009), which contained pre-established parameters, namely: the student's capacity for initiative and participation in lessons; questions asked by the students in relation to the content; curiosity in relation to the content; difficulties presented in relation to the lessons and the students' esteem in relation to the lessons (APPENDIX 2), for better organisation and clarity of the notes relevant to the work, both of the theoretical lesson worked on at the school and the theoretical-practical lesson proposed by the researcher.

On 07/10/14, the school's pedagogical coordinator was given the Free and Informed Consent form and the Consent for the Participation of the Person as a Subject (ANNEX 1), documents that legally supported and authorised the researcher to administer the questionnaire to the students.

The questionnaire used contained eight semi-structured questions (GERHARDT et al., 2009) (APPENDIX3). On that day, after authorisation from the pedagogical coordinator, the questionnaire was administered to two classes. Finally, on 10 October 2014, the same questionnaire was administered to the other two classes taking part in the research.

3.3 DATA ANALYSES

According to Dutra (2006, p. 8) "the comparative method consists of investigating things or facts and explaining them according to their similarities and differences".

In this context, we opted to analyse the data obtained using a qualitative-quantitative approach in the terms of Minayo (2008). With regard to qualitative analysis, Dalfovo; Lana and Silveira (2008, p. 9) point out that "[...] it is one that works predominantly with qualitative data, i.e. the information collected by the researcher is not expressed in numbers, or the numbers and the conclusions based on them play a minor role in the analysis".

On the other hand, in relation to quantitative analysis, the same authors emphasise that, "[...] its differential is the intention to guarantee the precision of the work carried out, leading to a result with little chance of distortion" (2008, p. 7).

Thus, in order to build a richer textual production for interpreting and validating the results, opting for a combination of quantitative and qualitative techniques makes the research stronger and reduces the problems of adopting only one of these groups of analyses (MINAYO, 2008). The aforementioned author goes on to say that combining different methods can make a significant contribution to enriching the analysis.

CHAPTER 4

STUDENTS' LEVEL OF INTEREST IN INTRODUCTORY PHYSICS TEACHING: APPROACHING CONTENT FROM DIFFERENT PERSPECTIVES

Here we will analyse the data collected using the observation "logbook" as the main tool, which was built up during the participant observations made in the theory-only classes taught by the school teacher and in the theory-practice classes taught by the researcher. This resulted in a set of two categories of analyses from which the data could be interpreted.

4.1 THE CLASSROOM SETTING FOR THEORETICAL CLASSES ONLY

This category analyses the data from the participant observations of the classes taught by the school teacher. To this end, it should be noted that during the two-week period, the teacher used three different ways to approach the content. These were: a) Taking the students to the school TV room to work on lessons using videos; b) Lectures; c) Promoting a debate.

At the beginning of the course, the teacher told the students that she was going to start the subject "Introduction to Physics". So, to give the students their first contact with the subject, the teacher took them to the school TV room to watch some videos.

The videos consisted of showing the students mostly science topics in general, i.e. involving content from the areas of biology, chemistry and physics. Each video lasted between 6 and 7 minutes on average. At class time, the students chose at random which one they wanted to watch, according to the decision of the majority.

In this sense, at the beginning of the lesson, the students were excited because the class was being held in the video room. However, during the course of the lesson, some showed tiredness or low self-esteem towards the class, not participating through questions or showing curiosity about the lessons. It should be noted that some of the videos did not present content related to Physics, such as the videos: Where do eggs come from?; Where does crying come from?; Where do shoes come from?

After the videos were played, during the breaks between them, it was noticeable that the teacher didn't problematise the content presented in order to allow the students to participate, or try to make a connection with some content that the students would see in the course of the subject, trying to obtain a diagnosis of their previous knowledge in relation to the subject, a factor mentioned as relevant in the study by Pereira and Padilha (2014). As a result, the students did not ask any questions or show any curiosity about the content covered. One of the students refers to one of the videos, A1TX: "not funny" in the middle of the video (where is the crying coming from?).

Pereira (2010, p. 1) mentions that "it is necessary to create a favourable environment for students to move from their non-scientific to their scientific conceptions". In this sense, there is a consensus that when students arrive at school they bring with them a baggage of knowledge built up through their social relationships, in other words, prior knowledge (AUSUBEL, 2003).

This highlighted the teacher's attempt to provide the students with a different working environment from the classroom. However, simply showing the videos without trying to stimulate students' opinions and participation in order to check their understanding of the subject can lead to a discouraging environment, as A1TX said.

Since, according to Pereira and Padilha (2014, p. 8), when they talk about the assumption that students have prior knowledge and this needs to be scientifically moulded when necessary, they point out, "we understand that in science teaching [...] this theory stands out as a basic element for building knowledge in students, especially scientific knowledge".

However, the teacher must try to create an environment that facilitates the students' learning and that can contribute to the construction of their scientific knowledge. After working on this first moment with the students, the teacher returned to the classroom to continue her work.

Before the lecture, the teacher asked the students in the class, based on the didactic videos shown in the previous lesson, what do you understand about physics? The students answered: A1TW: "exercise"; A2TW: "subject"; A1TY: "run"; A2TY: "play"; A3TY: "do exercises".

It can be seen that, in this first moment, the students associated the subject with physical activities. This lack of knowledge of the real objectives of studying physics may be associated with the fact that in previous grades, where the content should have been included in the science subject, teachers put more emphasis on studying biological sciences (biology)

(SOUZA et al., 2012), while in the 9th grade the subject in question is something new for students to learn.

In view of this, it can be seen that science teachers will face major challenges in providing 9th graders with a solid foundation in physics concepts in order to minimise the difficulties students encounter with this subject in secondary school.

After diagnosing the students' previous understanding of the subject's study objectives, the teacher copied on the board the new content to be worked on, the title of which was: Physical Quantities (Scalar and Vector).

In this sense, one of the students "fell asleep" while the teacher was copying the content on the blackboard. The teacher used the textbook as the main resource for this lesson.

During the explanation, the students chatted in parallel and didn't question the teacher about the content. During the lesson, the student who had been "sleeping" resumed his activities and started copying the second part of the content. While other students didn't copy the content, they didn't show any interest in the lesson.

After she had finished copying the subject on the board, the teacher asked the students to read the text briefly and explain what they understood about the reading, but not everyone read, they talked to each other. The teacher asked the class a question and one student answered, but he didn't fulfil the teacher's requirements.

It can be seen that, due to their first contact with the subject, the students had some difficulty answering the questions raised by the teacher. However, there are aspects that we should highlight in relation to her practice: trying to be clear in her definitions, trying to encourage student participation and reading.

Furman (2009, p. 9) points out that, "[...] what we see most often in schools", is the presentation of content to students as a ready-made substance, which must be "memorised" in order to solve exercises later, with the predominant reference being what is presented in the textbook.

Given this, this context clashes with what Carvalho et al. (2007) mention, that one of the goals of teaching is to create autonomous students who know how to think, criticise and make their own decisions.

On the other hand, it also goes against what Pereira and Padilha (2014) mention when they argue for a differentiated stance from the science teacher aimed at transformation and changing a reductionist view of the world for student learning.

So, in light of all this reflection, it's worth making it clear that we don't intend to hold science educators in the final grades responsible for the fact that physics is still approached from the "old practices" point of view, but we would like to point out that "in a society where scientific knowledge is overrated and technology is increasingly involved in everyday life [....]" (BRASIL, 1998) showing science in a more "lively" and dynamic way *is* essential in order to contribute to the formation of citizens who are aware of the current social situation and to stimulate students' taste for scientific knowledge in later grades.

After finalising the content on Physical Quantities, the teacher began teaching a new content, "Units of Measure". To get things started, the teacher organised a debate, which was worth 2 points. In this context, the teacher asked the class to split into two groups, one group with the topic "time" and the other with the topic "length".

The activity was carried out as follows: first the students were given 30 minutes to read and interpret the text, researching it in the textbook and discussing among themselves what they understood about the subject, as well as deciding who was going to present the content.

The text talked about the historical development of these two units of measurement, from the earliest attempts to the most current ones that man uses to measure these quantities. After the reading, the groups came face to face to hold a debate.

When the groups socialised, it was noticeable that in the time group, one of the students showed knowledge and the ability to interpret and argue about the text. On the other hand, most of the students in this group found it difficult to follow her reasoning. In the length group, there was less discussion and socialisation of the content between the students.

It should be noted that the "time" group discussed other scientific theories (big bang) and one of the students asked, what is a big bang? One student replied, "It was an explosion that happened a long time ago where 98 per cent of matter was formed". Here, the students' previous knowledge of other science content was highlighted and, although the student's answer is not in total agreement with the literature, it is up to the teacher to enhance and transform this spontaneous knowledge into scientific knowledge.

Thus, during the course of the lesson, this group managed to discuss the content among themselves, while the "length" group showed less curiosity and questioning about the content, i.e. they were less organised than the other group. It was noted that in the length group, "one of the students" was distant from the discussion and away from the group, showing a lack of interest in the lesson.

It should be noted that, according to Pereira and Padilha (2014), each student brings a different baggage of knowledge, meaning that depending on the student's knowledge structure, they may find it easier or more difficult to learn a particular subject. In this way, we can infer that each student has a different time and way of learning, which we believe is important for the teacher to recognise this characteristic, in order to provide learning that reflects the majority of students.

In this sense, the time group started the debate, and one of the students clearly explained a preamble about the history of time according to the readings in the textbook. The student showed a good ability to argue and interpret the text. As for the length group, the student showed some difficulty in arguing about the topic, as she said, "I didn't study, ... it's .... harn"

It was also observed that the time group worked as a team, where the majority tried to contribute to the debate, passing the book on to each other, reading and talking in each other's ears to get someone to contribute. The length group, on the other hand, worked in a less collective way, and most of them kept to themselves, showing no curiosity or questions about the debate.

The teacher said, "Come on, the length group is very quiet". The teacher instigated the students, and this made them participate more in the debate, give their opinions and put forward their ideas.

In this context, it was noted that the students were gradually familiarising themselves with quantities and their units of measurement, working as a conceptual evolution[1] (PACHECO, 1997).

It should be noted that the proposed activity motivated the students, either because of

---

[1]According to Pozo and Crespo (2009), conceptual evolution is the process by which the student abandons their previous knowledge when the content they are working on becomes meaningful and later gives way to the appropriation of scientific knowledge.

the competition, the nature of the work or because it was an activity that would earn points. It should also be pointed out that some physical concepts are more difficult to work on with experiments, and so it was found that activities of this nature motivated, instigated and allowed the students to participate and interact more in class, becoming active agents in the construction of knowledge.

Another aspect that stands out is that the teacher emphasised the history of science in the lesson. However, after the end of the activities it was noticeable that the teacher concluded the lesson satisfied and the students euphoric about the competition, thus changing the classroom scenario that had been established in previous moments.

On the other hand, in order to make lessons more attractive and more dialogical with students, Furman (2009) suggests that science teachers start by illustrating the lesson with an everyday situation and when students understand what is being exemplified in their daily lives, it is a good time for the teacher to give them the scientific conceptualisation of the thing.

The author also emphasises that respecting this didactic sequence, i.e. the connection between everyday examples and scientific phenomena, does not necessarily require working with practical lessons, but it would be enough for the students to associate the examples shown by the teacher with scientific literature. Thus, we observed that the debate organised by the teacher and the precepts of the author above were somewhat different from the stance which values the direct definition of concepts on the blackboard and the students listening passively.

4.1.2 THE CLASSROOM SCENARIO IN THEORETICAL-PRACTICAL CLASSES

The theoretical-practical classes, taught by the researcher, were held in two timetables and consisted of the following: first, the theoretical content was covered using slides, highlighting various everyday situations involving the subject in question, and then the experiments were carried out with the students.

The students were then given a problem, such as why the toothpick was higher than the dice in the mini-catapult experiment. They were then asked to carry out the experiment and then answer the questions raised by the researcher, supported by the theory previously explained in class and by the presence of the researcher providing theoretical and technical support to the students during the activity.

To this end, we started from the assumption made by Carvalho et al. (2007), who point out that the problem generates the promotion of varied attitudes in students, in other words, it motivates, instigates, challenges and promotes greater student involvement in classes. It should also be emphasised that this was based on the principle mentioned by Arruda and Laburú (2009), who classify the type of approach against the backdrop of experimental situations that the teacher can develop in the classroom, taking into account the student's level of cognitive development.

Thus, the students were given a problem according to Carvalho et al. (2007) and the experiments followed a demonstrative type classification (ARRUDA; LABURÚ, 2009), because the students were from the 9th grade and would have to solve the problem and, at the same time, check whether the theory matched the practice, based on the execution of the experiments.

In fact, it turned out that the students were interested, interacting through questions and actively participating in the lessons. It was also noticed that they participated enthusiastically in the experiments, were very curious about the content, challenging the researcher with further questions about the concepts covered, and most showed an interest in taking part in the experiments.

In view of this, Arruda and Laburú (2009, p. 65) also emphasise that "the main aim of this activity would be to reach a large number of students by carrying out interesting experiments that would arouse their attention to science". In this way, it was noted that the classes responded to the questions raised by the teacher, since the problem situations provided greater interactivity and students' attention to the lesson.

On the other hand, it should also be noted that after the theoretical-practical lessons, three questions were worked on with the students involving the mathematisation of concepts (APPENDIX 1), due to the fact that Physics is a science that not only explains various phenomena qualitatively, through principles, but also requires measurement procedures, i.e. a quantitative approach (LABURÚ; MOURA; SALES, 2010).

However, as the concepts of Newton's second law were worked out in conjunction with the mini-catapult experiment, this law can be expressed by the following mathematical expression: **F = m.a** Where **F** is the force applied to a given body, and its respective unit of measurement is the Newton (N), **m** is the mass of the body and its respective unit of measurement is the Kilogram (Kg) and **a** is the acceleration acquired by the body and its respective unit of measurement is the metre per second squared ($m/s^{2)}$.

After working on the experiments, it was noticeable that the students were enthusiastic and duly copied down the questions and then tried to solve them. Thus, classes X, Y and W showed interest in the activity and most of them managed to solve the questions on their own. However, class Z showed no interest in solving the activity. At first, some of the students in class Z said "ah, now with maths I don't think Physics is cool anymore", showing that the mathematical approach to Physics becomes something adverse to learning the subject.

Santos; Gomes and Praxedes (2010, p. 5) point out that one of the difficulties that hinder many students in the teaching of Physics is, "(...) deficiency in basic knowledge of Maths. This factor hinders students in learning this subject".

It should be noted that the quantitative approach, i.e. the use of measurement procedures in learning physics, is yet another obstacle that science teachers need to confront in the process of teaching and learning this subject. It was only after the help and encouragement of the researcher that some of the students in Z managed to solve the proposed questions.

In this sense, there are additional factors, in addition to those described here, that make it difficult for students to learn physics, as indicated by Pozo and Gomes Crespo (2009), factors that do not need to be explored in greater depth for the purposes of this study. As a result, it was found that experimental activities associated with the attitude of a differentiated teacher, i.e. one who is concerned about the students' learning and goes to great lengths to try to overcome the challenges of school reality, can be a way of trying to bring students' taste for science closer together, based on the scenario established at the time of the theoretical-practical lessons, differing from that environment in which the students just listen and the teacher lectures.

CHAPTER 5

REFLECTIONS ON THE LEVEL OF STUDENT INTEREST IN INTRODUCTORY PHYSICS TEACHING: THEORETICAL AND PRACTICAL CONTENT

This chapter will analyse the data collected using a questionnaire containing eight semi-structured questions (GERHARDT et al., 2009) applied to the students in the four classes. In questions 1 to 3, it is important to bear in mind that the assessment should have concerned the subject teacher's theoretical lessons. However, it was noted that the students evaluated and compared the methodology used in the intervention with the lectures given by the teacher in question.

It's worth remembering that the four classes surveyed were named X, Y, Z and W. The students' answers were organised in alphabetical order. Students were given codes consisting of the letter A and its respective order, and T referring to the class as designated above. For example, student 1 from class X will be labelled A1TX. The same rule was applied to the other answers analysed.

In this sense, the decision was made to develop categories of analysis as mentioned by Bardin (2011), since many of the answers converged. Thus, all the answers were pre-selected, analysed and grouped according to each established category (BARDIN, 2011). The results will be analysed below.

In the first question, **47** answers were analysed from the students in the four classes that took part in the survey, according to the categories below expressed in the graph in relation to the percentage of student answers.

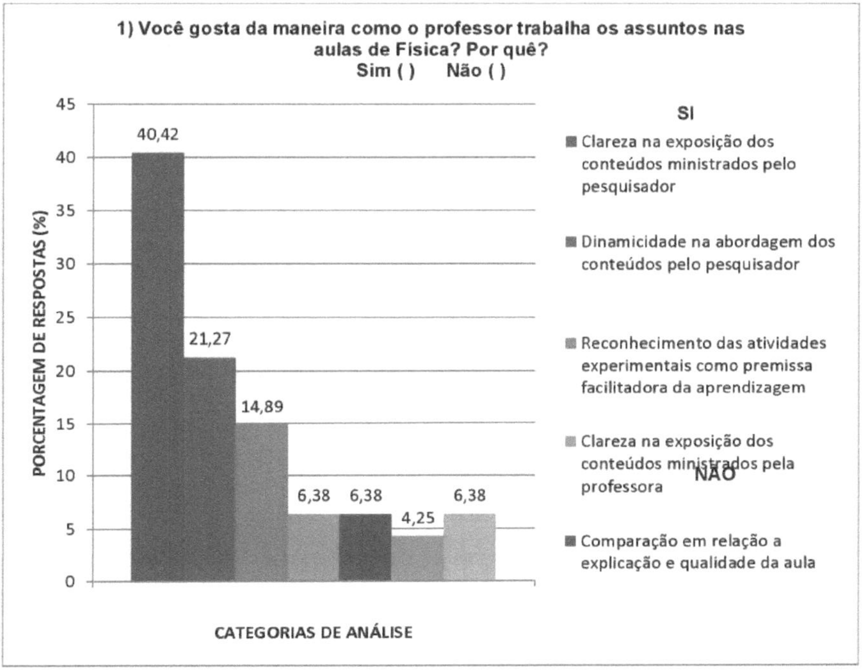

Figure 4 - Graph of the first question of the questionnaire applied to students. **SOURCE**: Data collection instrument.

Of the total responses analysed, 10 were grouped into the category "dynamic approach to the content", 02 into the category "difficulties in understanding the content taught by the teacher", 03 into the category "approach used by the teacher as a demotivating factor for learning the content", 19 into the category "clarity in the presentation of the content taught by the researcher", 03 into the category "clarity in the presentation of the content taught by the teacher", 03 into the category "comparison in relation to the explanation and the quality of the lesson" and 07 grouped into the category "recognition of experimental activities as a premise for facilitating learning".

After the quantitative apparatus, the students' answers will be analysed according to each category established, following the same pattern for the other questions. To this end, tables were drawn up expressing the title of the category of analysis and the students' responses, in order to better organise and discuss the results found.

It is worth emphasising that the pronoun **"he"** refers to the students' assessment of the lesson taught by the researcher and the pronoun **"she"** refers to the students' assessment of the introductory physics lessons taught by the school teacher. The tables derived from the analyses of the students' responses are shown below.

CATEGORY OF ANALYSIS	STUDENTS' ANSWERS
DYNAMIC APPROACH TO CONTENT	A2TX: "He works on the subjects in a dynamic way and makes it easy to learn." A7TY: "Yes, because he knows how to explain, and the lessons were more interactive." A9TZ: "Yes, because most of the time he plays games and it's really nice to learn like that." A14TW: "Yes, because he works in such a way that everyone can participate."

Table 1 - Analysis category A of the first question. **Source:** Data collection instrument.

As can be seen from the answers in Table 1, the literature already points out that didactic strategies involving experimentation can be relevant mechanisms for presenting students with a more lively and dynamic science, thus clashing with the practice adopted by the predominant use of passing on information (ATAIDE; SILVA, 2011).

In this sense, the teacher adopts a different approach to scientific concepts, providing students with more flexible and dynamic learning moments, in which, when carrying out an experimental practice, students become euphoric and start to develop a taste for scientific knowledge.

Imbernon et al. (2009, p. 87) point out that "dialogue through experimentation and interactivity can be the first step to achieving student participation in the educational process [,,,]". In this case, when students' opinions are valued and their participation in class is stimulated, they will feel more involved in the teaching and learning process (COELHO; NUNES, 2003).

The same authors also point out that experimental activities encourage students not to remain solely in the world of concepts or abstractions. Thus, it can be said that through experimental activities the student sees himself more clearly involved in the construction of knowledge, thus reflecting greater dynamism in the approach to concepts in class.

CATEGORY OF ANALYSIS	STUDENTS' ANSWERS

DIFFICULTIES OF UNDERSTANDING OF THE CONTENT TAUGHT BY THE TEACHER	A5TX: "No, sometimes the content is difficult to understand even if I explain it several times." A1TY: "no, because the teacher says some words that I don't understand, so it's hard to understand"

Table 2 - Category of analysis B of the first question **Source:** Data collection instrument.

This category, shown in Table 2, highlights the difficulty students have in assimilating the concepts covered by the teacher. In fact, according to Santos; Gomes and Praxedes (2010, p. 4), when they talk about the concepts included in the curricular component in question, they mention that "many have this subject in mind as something impossible to learn [...]".

However, it should be emphasised that physics is an experimental science that is widely applied in everyday life. In this way, clarity in the teacher's exposition of the concepts and the attempt to offer experimentation to facilitate understanding of the content in class can make a significant contribution to learning and to demystifying the idea that the subject of Physics is impossible to learn.

CATEGORY OF ANALYSIS	STUDENTS' ANSWERS
APPROACH USED BY THE TEACHER AS A DEMOTIVATING FACTOR FOR LEARNING THE CONTENT	A6TX: "no, because it's the way she approaches her lessons" A6TY: "no, because I don't find it very interesting." A16TY: "no, I don't think it's very nice, it would be nice if the content was explained more."

Chart 3 - Analysis category C of the first question. **Source:** Data collection instrument.

In this category, shown in table 3, it can be seen that the students evaluated the teacher's approach to the content as an adverse factor in their learning. According to Souza and Heineck (2006, p. 7) "The teacher's attitude towards their students can determine their interest in the content developed, and everything indicates that interest can be provoked by the teacher; [...]". Also according to the authors, it is important for the teacher in the classroom to contextualise the content, providing students with moments to relate what they learn in class to what they observe in everyday life.

In this way, the teacher is challenged to show the students how relevant the content is and to convince them that many of the physical principles studied guide their daily lives. Zompero; Passos and Carvalho (2012, p. 46) point out that "it is necessary to bring moments of discovery into the classroom that make sense to the student, that are real and challenging problems, so that they feel like reflecting on what they are investigating". Thus, it is believed that learning moments that allow students to go beyond the "mere reception of concepts" can favour students' liking for science classes.

CATEGORY OF ANALYSIS	STUDENTS' ANSWERS
CLARITY IN THE PRESENTATION OF THE CONTENT GIVEN BY THE RESEARCHER	A13TX: "Yes, because he works on subjects that my friends and I can understand" A17TY: "Yes, because the way he explains, we understand better and go deeper into the subjects." A5TZ: "Yes, because he explains the subjects well [...]". A19TW: "Yes, for his good methods of teaching physics."
CLARITY IN THE TEACHER'S PRESENTATION OF THE CONTENT	AI 1TY: "Yes, because when she gives her lesson she shows several examples and in a very nice way." AI 8TY: "Yes, because in different ways the teacher brings knowledge, which are creative lessons that make us understand physics even more" A21TW: "Yes, because she explains things well when we don't understand, she explains them several times, it's very nice."
COMPARISON OF THE EXPLANATION AND THE QUALITY OF THE LESSON	A8TX: "Yes, because he explains it much better than the teacher." A11TX: "Yes, because the way he explains it is better for our learning." A1TW: "yes, because he explains well and is better than the teacher"

Table 4 - Categories of analysis D, E, F of the first question.
Source: Data collection instrument.

Another factor observed in table 4 was the relevance of the clarity of the content taught by the researcher for the students' learning. Silva (2009, p. 7), when talking about the teaching profession, tells us that "this is one of the professionals who most needs to keep up to date, combining the task of teaching with the task of studying [...]". In view of this, working on science content in a way that favours students' learning will require the teacher to take time to plan, as well as a constant dedication to reading in order to keep up to date.

On the other hand, Pacheco (1997) explains that when teachers recognise that their work in the classroom goes beyond merely passing on information, their work will be doubled. By recognising this, the teacher will go to great lengths to find activities that make their lessons more comprehensible to the students, such as adding experimental activities to their theoretical lessons.

For example, Santos; Gomes and Praxedes (2010, p. 6) explain that "practical or experimental classes are of great cognitive value. This is based on the behaviourist approach, which assumes that learning takes place through experiences". Thus, in the proposal to work with practical activities, lessons become differentiated, enabling greater proximity between teacher and student in the construction of knowledge.

Pereira (2010, p.6) notes the demands of the profession when working with teaching methods that favour experimentation, "[...] the teacher who proposes an investigative activity must, in addition to knowing the subject they are teaching, become a questioning, argumentative and challenging teacher, guiding the teaching process".

In addition, teachers need to be prepared and recognise that the "content mastery" factor in science classes is a prerequisite for those who want to contribute to the construction of meaningful learning[2] for their students.

CATEGORY OF ANALYSIS	STUDENTS' ANSWERS
Recognising experimental activities as a premise for facilitating learning	A15TY: "Yes, I like it, because it's a different way of working." A19TY: "yes, because the way the teacher gives the lesson is very creative, and also talking about the practical lessons." A9TW: "(...) in practical lessons it's easier to understand." A11TW: "Yes, because he does experiments and it's easier to understand the subject."

Table 5 - Analysis category G of the first question. **Source:** Data collection instrument.

Theorists such as Silva (2009); Medeiros; Leite and Oliveira (2007), Galiazzi et al., (2001) point out that experimental activities, when used in the process of teaching and

[2]In the view of Pereira and Padilha (2014), learning becomes more meaningful the more of a meaningful relationship the student is able to form between what they already know and the new content that will be provided as an object of learning.

learning science, can be considered a viable alternative for minimising the difficulties of learning and even teaching physics in a more consistent way.

Moreira and Penido (2009, p. 2) note that "[...] historically, various attempts to improve the quality of physics teaching have been based on experimental activities". Thus, still in the terms of the aforementioned authors, it is noted that:

"[...] the scientific community has been working on the production of didactic experiments, both with sophisticated materials and with low-cost or recyclable and recycled materials" (MOREIRA; PENIDO, 2009, p. 3), thus seeking to minimise the difficulties encountered by students, both in primary and secondary schools, in the process of teaching and learning the curricular component in question.

Borges (2002), in his work on evaluating the use of teaching laboratories in science teaching, explains that experimental activities can..,
in some cases, take on a traditional character when employed, with the backdrop being the mere execution of ready-made "recipes" for the students. Still within the author's parameters, in order to achieve the goal of facilitating learning, it is recommended that practical activities focus only on the desired aspects and the student's level of cognitive development, with careful planning that takes into account the students' previous ideas in Ausubel's terms (2003).

On the other hand, Araújo and Abib (2003) agree that the use of experimental activities associated with theoretical physics lessons has been pointed out by teachers and students as one of the relevant teaching strategies for trying to minimise the adverse factors in learning and teaching physics in a meaningful and consistent way. In view of this, the data expressed in table 5 and the literature lead us to realise that experimentation associated with science teaching seems to be one of the most fruitful ways of combating the obstacles that teaching this subject imposes on teachers.

In the second question, **41** answers were analysed from the students in the four classes that took part in the survey, according to the categories below expressed in the graph in relation to the percentage of student answers.

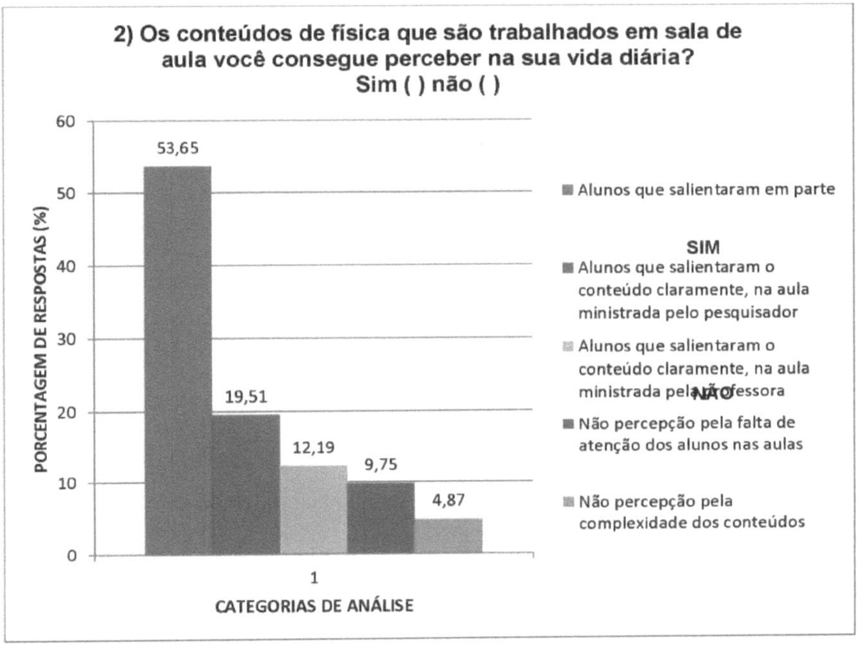

Figura 5 - Graph of the second question of the questionnaire applied to students. **Source**: Data collection instrument.

Of the total number of responses analysed, 22 were grouped into the category "students who emphasised the content in part", 08 into the category "students who emphasised the content clearly in the class taught by the researcher", 05 into the category "students who emphasised the content clearly in the class taught by the teacher", 04 into the category "no perception due to the students' lack of attention in class" and 02 were grouped into the category "no perception due to the complexity of the content".

CATEGORY OF ANALYSIS	STUDENTS' ANSWERS
STUDENTS WHO CLEARLY EMPHASISED THE CONTENT IN THE LESSON TAUGHT BY THE RESEARCHER.	A14TX: "yes, because in the case of a car, if you don't wear a seatbelt, you'll be thrown out of the car." A18TY: "yes, because physics is present everywhere, in my day-to-day life, speed, force, (...)" A4TZ: "Yes, because they showed examples from everyday life, like when a car crashes." A3TW: "yes, especially inertia and gravity, which are my favourites"

STUDENTS WHO EMPHASISED THE CONTENT CLEARLY, IN THE CLASS TAUGHT BY TEACHER	A14TY: "Yes, because what can be measured is a physical quantity, like doing work in our daily lives." A18TY: "(...) day and night, etc., important concepts." A2TW: "yes, like time on the clock" A21TW: "Yes, a little, when we talk about time, length."

Chart 6 - Categories of analysis A, B of the second question. **Source:** Data collection instrument.

It is noteworthy that some of the students clearly emphasised the content worked on by the researcher, eight of them, and five of them by the teacher, as shown in Table 6. According to Pereira and Padilha (2014, p. 8) "learners only learn meaningfully if they are able to bring knowledge into their own reality".

On the other hand, the same authors also emphasise that these students bring with them to school prior knowledge of the content being covered, which refers to knowledge built up in day-to-day life and in individuals' social relationships. In this way, students' prior knowledge needs to be taken into account and re-signified as scientific knowledge at school.

To this end, Tauceda; Nunes and Del Pino (2011, p. 134) mention that experimental activities can help to bring science teaching closer to the reality of students, "[...] in order to improve the acquisition of knowledge and the mental development of students, by solving problem questions that value their previous concepts and rationalisations".

Thus, it is argued that experimentation activities would be an alternative for students to realise concepts in practice, based on problem solving, and relate the resolution of these problems, with the help of the teacher, to their day-to-day lives, so that, in this way, classroom content gains greater significance in the teaching and learning process.

CATEGORY OF ANALYSIS	STUDENTS' ANSWERS
STUDENTS WHO PARTLY EMPHASISED	A2TX: "Yes, the way he explains it is easy to fix in the mind, so it becomes easier to understand on a day-to-day basis." A1TY: "yes, because the teachers usually give their explanations and work according to our day-to-day lives, which is nice for me." A8TZ: "yes, many things I see in class happen at home and this is part of my life and I learn more." A18TW: "yes, whether we like it or not, physics is present in our lives in everything"

Table 7 - Analysis category C of the second question.
Source: Data collection instrument.

A large proportion of the students, **22 of** whom were identified, did not clearly define any content related to their daily lives, as shown in table 7. In this context, in science teaching, as Ausubel (2003) mentions, there is a consensus that learning based on simple memorisation does not lead students to transform their reality and enhance their creativity.

Pereira and Padilha (2014, p. 23), when mentioning the construction of knowledge in science, point out that "[...] it is essential that the content worked on in school is based on students' everyday situations, so that scientific knowledge is not imposed as 'true', immutable, crystallised". In this way, it can be inferred that for science teaching to become more attractive to learners, it is important to mention that students know how to transpose what they learn at school into reality, with the aim of understanding it better in order to transform it.

CATEGORY OF ANALYSIS	STUDENTS' ANSWERS
LACK OF PERCEPTION DUE TO STUDENTS' LACK OF ATTENTION IN CLASS	A7TY: "no, because I don't pay attention". A4TW: "no, because I don't pay attention" A10TW: "Sometimes I can't understand because I don't pay attention." A12TW: "Not much, because I have to pay more attention."

Table 8 - Analysis category D of the second question.
Source: Data collection instrument

In this category, shown in table 8, these students recognise that the factor "lack of attention in class" contributes to them not perceiving the content worked on in class in relation to their daily lives. In this sense, Buch and Schroeder (2013, p. 73) draw attention to the following excerpt, "[...] with the evolution of humanity there has been an overvaluation of scientific and technological knowledge and, today, it is not possible to think of the individual without having at least a basic knowledge of scientific knowledge".

In this way, it can be seen that science teachers still have the task of attracting students' attention and the challenge of making them realise, in the course of their lessons, that scientific concepts are relevant to students' lives and to building their full citizenship.

CATEGORY OF ANALYSIS	STUDENTS' ANSWERS
NO PERCEPTION OF THE	A2TZ: "some things, because others aren't very easy to understand." A7TZ: "some

COMPLEXITY OF THE CONTENT	things, because others aren't very easy to understand."

Chart 9 - Analysis category E of the second question. **Source:** Data collection instrument.

Table 9 shows that the "complexity of the content" factor proved to be an obstacle to students associating the content worked on with their reality. In fact, according to Pozo and Gomez Crespo (2009, p. 191), when they point out the difficulties that some students have in understanding certain physics content, it requires the individual to "[...] resort to idealised and simplified representations that are far removed from reality, or at least from what we perceive as our reality".

It is because of this that many students find it difficult to associate the phenomena that physics explains with their lived reality. This leaves the reflection that the teacher needs to look for ways in which the student can find meaning in learning the curricular component in question, to try to minimise the complexity of "idealisation" that many have as an adverse factor in their learning.

In the third question, **49** answers were analysed from all the students in the four classes. They were grouped according to the categories below expressed in the graph in relation to the percentage of student responses.

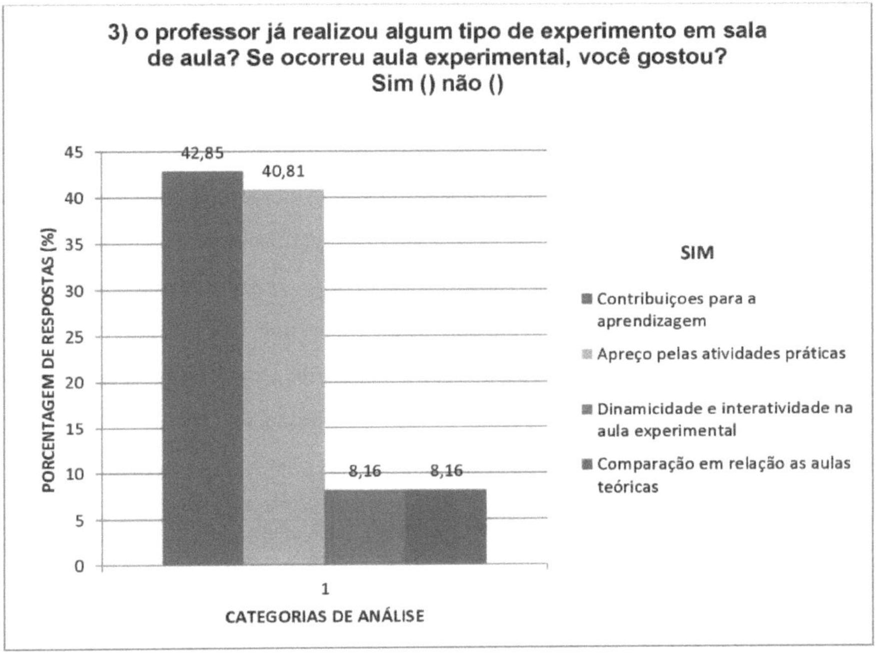

Figura 6 - Graph of the third question of the questionnaire applied to students. **Source**: Data collection instrument.

Of all the answers analysed, 4 were grouped into the category "dynamism and interactivity in the experimental class", 21 into the category "contributions to learning", 20 into the category "appreciation of practical activities" and 4 were grouped into the category "comparison with theoretical classes".

CATEGORY OF ANALYSIS	STUDENTS' ANSWERS
DYNAMISM AND INTERACTIVITY IN THE EXPERIMENTAL CLASSROOM	A2TX: "Yes, he explained it in a simple and dynamic way." A4TX: "Yes, it was funny, dynamic, everyone participated and I understood." A14TX: "Yes, everyone interacted in his class." A8TY: "yes, in practice it's nice and fun (...)"

Table 10 - Analysis category A of the third question.
Source: Data collection instrument.

Pereira (2010, p. 3) points out that the role of experimental teaching, when associated with science lessons, "is related to the awareness of the need for teachers to adopt a different approach to teaching and learning science". According to the author, students are encouraged to participate and the teacher's pedagogical help is essential to the interactive and dynamic processes that characterise experimental practice in science.

According to Araújo and Abib (2003, p. 190), when they talk about the efficiency of experimental classes as didactic strategies in science teaching, they allude to the fact that such activities have the "ability to stimulate the active participation of students [...]".

Given this assumption, we believe that the teacher's pedagogical intention in inviting students to participate in the construction of knowledge, respecting their previous opinions in relation to the concepts, can be a relevant factor in changing the way the subject is presented to students, thus favouring greater dynamism in classes. According to the data shown in chart 10.

CATEGORY OF ANALYSIS	STUDENTS' ANSWERS
CONTRIBUTIONS TO LEARNING	A12TX: "Yes, because it's better for learning." A3TY: "Yes, I really enjoyed it. It was nice to learn from the experiments and exercises." A3TZ: "Yes, I thought it was very cool because we learn faster." A11TW: "Yes, because it's easier to understand the subject."

Table 11 - Analysis category B of the third question.
Source: Data collection instrument.

In this category, shown in chart 11, the students listed that they liked the experimental lesson because it contributed to their learning. According to Carvalho et al. (2007, p. 35), "teaching takes place through the constructive mental activity of the student, who manipulates, explores, performs, reads, asks questions and expresses their ideas".

Therefore, offering students the opportunity to manipulate reality and verify the materialisation of theoretical concepts when carrying out practical activities added to students' learning, according to the data shown in table 11 - analysis category B. It is therefore favourable to point out that experimental activities can contribute to greater student involvement and the perception that theory associated with practice has a positive influence on the way students construct knowledge.

CATEGORY OF ANALYSIS	STUDENTS' ANSWERS

APPRECIATION FOR PRACTICAL ACTIVITIES	A11TX: "Yes, the catapult, which was a cool experiment, the lighter and higher the object is launched." A15TY: "Yes, I really liked it and I want you to repeat it more often." A4TZ: "Yes, I liked the inclusion of the students in the experiments." A18TW: "Yes, I liked it, it taught us a lot about physics."

Chart 12 - Analysis category C of the third question. **Source:** Data collection instrument.

It is necessary to infer that the literature points out that "[...] the new requirements for science teaching include procedural and attitudinal dimensions in addition to the conceptual dimension" (PEREIRA; RIBEIRO; FREITAS, 2014, p. 15). In this sense, according to Carvalho (2004), when science teaching is based on these dimensions, it gives way to the transition from the concept of pure science to the concept of Science, Technology and Society (STS).

Therefore, from the author's point of view, we should no longer pass on neutral, ready-made science concepts to future generations, while today we can no longer think of teaching science without linking it to discussions about the technological and social effects that this science has on moulding today's society.

In this sense, from an attitudinal point of view, it is up to the teacher to direct their teaching towards a broader cultural purpose that will help students develop the skills to make informed and critical decisions about the role that science and technology play in the development of society. The aim is to make the teaching of this subject more attractive to the student's eye and, on the other hand, to avoid the mere transmission of information.

In this context, Pereira (2010, p. 6) mentions the behavioural change between students and teachers when an experimental lesson is proposed in the teaching and learning process: "[...] The student is no longer just an observer of the generally expository lessons, but starts to argue, think, act, interfere and question".

Thus, when these students are invited to interact in class and understand the world around them based on the physical principles presented and consolidated in the experimental representation in an attempt to combat the teaching of physics that has been carried out through the presentation of concepts, laws and mathematical formulas, repetitive exercises that only stimulate memorisation, students tend to develop a greater appreciation for the subject (SANTOS; GOMES; PRAXEDES, 2010).

On the other hand, it is recognised that physics is a science that inevitably explores a quantitative approach based on measurement procedures. However, combining the quantitative approach, which is generally used to show students mathematical exercises, with the qualitative approach, working on physical concepts and principles through experiments, would be a way of adding to the students' meaningful learning by offering them the conditions to develop a greater affinity with the subject.

Coelho and Nunes (2003, p. 30) see experimentation as a way of "[...] favouring the establishment of a link between the world of objects, the world of concepts, laws and theories and that of symbolic languages".

In this way, when the quantitative nature of physics teaching - the mathematical approach - and the qualitative nature - the exposition of principles - are associated, i.e. working with theory and practice in lessons, according to the data shown in table 12, the students showed greater involvement and interest in more lessons in the science subject.

CATEGORY OF ANALYSIS	STUDENTS' ANSWERS
COMPARISON WITH LECTURES	A15TX: "Yes, because I could see it in practice and in theory." A16TX: "Yes, very much so, not least because the methodology was better than what we're used to seeing." A12TY: "Yes, because we don't always understand what we've seen in books and we understand it better when we do it." A6TZ: "Yes, because it was better."

Table 13 - Analysis category D of the third question.
Source: Data collection instrument.

Theorists such as Pereira (2010); Santos; Gomes and Praxedes (2010), when talking about the teaching of Physics, draw attention to the way in which the teaching of this subject is generally presented to students, as a teaching carried out through the presentation of concepts, laws, mathematical formulas, repetitive exercises that are ready through books, handouts that only stimulate memorisation.

On the other hand, according to Borges (2003, p. 293), it is necessary to emphasise that "first and foremost it is necessary to emphasise that teaching, not just science, is a complex and problematic activity". Thus, making science lessons more attractive to students by confronting the model of "passing on information", such as associating theoretical lessons with experimentation, is no easy task.

It is therefore believed that the use of experimental models for teaching science is recognised as an interesting alternative for building knowledge, says Pereira (2010). Silva and Rocha Filho (2010) point out that experimental activities can play a role in proving theory. This activity can establish a positive relationship in the approach to content, because according to the students' responses expressed in table 13, experimentation was a different type of approach to the classes they were used to seeing. It therefore had a relevant impact on the way knowledge was constructed.

In questions 4 to 8 of the questionnaire, the students had to evaluate the methodology used by the researcher in the intervention. It was noted that some students also compared the lessons given by the researcher to those given by the teacher.

In the fourth question, **54** answers were analysed from all the students in the four classes. They were grouped according to the categories below expressed in the graph in relation to the percentage of student responses.

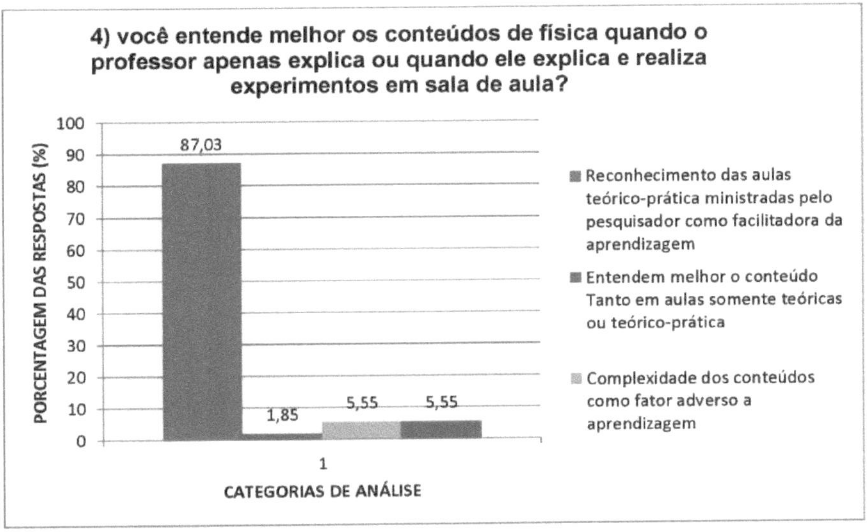

Figura 7 - Graph of the fourth question of the questionnaire applied to the students.
Source: Data collection instrument.

Of the total number of responses analysed, 47 were grouped into the category "recognition of the theoretical-practical classes given by the researcher as facilitating learning", 01 into the category "better understanding of the content in both theoretical and theoretical-practical classes", **03** into the category "complexity of the content as an adverse factor to learning" and **03** were classified in the category "comparison in relation to the explanation and quality of the class".

CATEGORY OF ANALYSIS	STUDENTS' ANSWERS
RECOGNISING THE THEORETICAL-PRACTICAL CLASSES TAUGHT BY THE RESEARCHER AS A FACILITATOR OF LEARNING	A5TX: "When he explains and carries out experiments, because we can better understand what the subject is about." A13TY: "I understand the content better when the teacher explains it and carries out experiments because it's better to understand." A5TZ: "It's easier to understand when he explains and does experiments, it helps the student better in his lessons." A3TW: "carry out experiments! Because I think we can understand better in practice."
BETTER UNDERSTAND THE CONTENT IN THEORETICAL OR THEORETICAL-PRACTICAL CLASSES ONLY	A2TX: "it doesn't matter, any of the ways the teacher approaches the subject is fine."

Chart 14 - Category of analysis A, B of the fourth question. **Source:** Data collection instrument.

It is recognised that theory is essential in science teaching, but Coelho and Nunes (2003, p.37) point out that "relying on theory in favour of practice and experimentation is also an excellent way of learning it". Still according to the authors, "through practical work and experimental activities, students should realise that in order to uncover a phenomenon, a theory is necessary" (COELHO; NUNES, 2003, p. 38).

Thus, in accordance with the literature and the students' responses, expressed in chart 14 - grouped in category A, theoretical-practical classes made it possible for what is preached in theory or in the textbook to become clearer and more comprehensible to the student's cognitive field, that is, when the work methodology adopted by the teacher in class allows theory and practice to go hand in hand.

On the other hand, it was noted that one of the students answered that both the methodology chosen by the teacher in class is relevant to their learning, see table 14 - category B, thus showing that each student learns according to their cognitive structure and capacity, in the terms of Pereira and Padilha (2014). However, attention should be drawn to the high number of students who emphasised that practical activities provided better learning conditions.

CATEGORY OF ANALYSIS	STUDENTS' ANSWERS
COMPLEXITY OF CONTENT AS AN ADVERSE FACTOR TO LEARNING	A10TY: "More or less, because some things are more difficult for one and others." A12TY: "More or less, because I'm not very good at physics." A14TY: "No, because I don't really understand."

Chart 15 - Category of analysis C of the fourth question. **Source:** Data collection instrument.

Lima et al. (2009) point out that due to the complexity of physical concepts, most students generally find it difficult to understand them simply by being exposed to theoretical descriptions.

On the other hand, Oliveira et al. (2010, p. 32) point out that "experimental activities can help to deepen the theoretical aspects of physics". According to the authors, "[...] simply carrying out experimental activities does not guarantee learning [...]" (OLIVEIRA et al., 2010, p.32).

Having said that, in order to add to and provide conditions for students to build their knowledge in Physics, it is essential that the teacher, in addition to proposing experimental classes, should theoretically subsidise the student in order to overcome the cognitive obstacles they may come across in assimilating the concepts.

CATEGORY OF ANALYSIS	STUDENTS' ANSWERS
COMPARISON OF THE EXPLANATION AND THE QUALITY OF THE LESSON	A17TY: "Yes, because he explains very well, and his lessons are very nice." A5TW: "When he explains it, it's better, it's more detailed." A12TW: "I think experiment activities are better because they make lessons better."

Table 16- Analysis category D of the fourth question.
Source: Data collection instrument.

Agostini and Delizoicov (2009, p. 4) point out that "didactic experimentation needs to be understood as an activity that, together with others, aims to find an answer to a proposed problem." These authors make us think that in addition to trying to associate experimentation in the teaching and learning process. In other words, a differentiated attitude on the part of science educators that goes beyond the mere transmission of knowledge.

Teachers need to have other premises, such as mastery of the content being taught in order to provide greater clarity in their explanations, planning of activities and dialogue in the construction of knowledge. Therefore, by adopting this type of attitude in the classroom, the teacher, as shown in Table 16, can add significantly to bringing students closer to the subject.

In the fifth question, **39** answers were analysed from all the students in the four classes. They were grouped according to the categories below expressed in the graph in relation to the percentage of student responses.

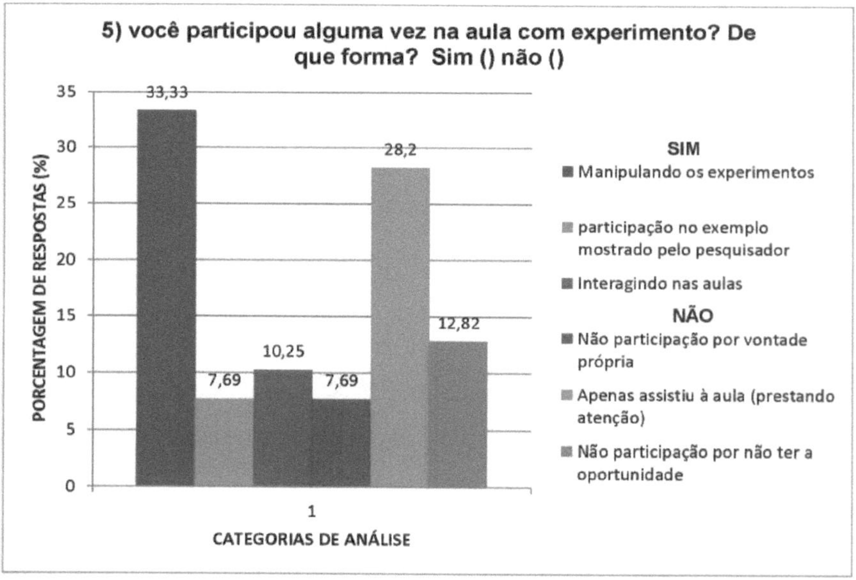

Figura 8 Graph of the fifth question of the questionnaire applied to the students. SOURCE: Data collection instrument.

Of all the answers analysed, 4 were grouped in the category "interacting in the classes taught by the researcher", 13 in the category "manipulating the experiments", 3 in the category "participating in the example and/or activity worked on by the researcher", 11 in the category "just watching the class (paying attention)", 3 in the category "not participating of their own free will" and 5 in the category "not participating because they didn't have the opportunity".

CATEGORY OF ANALYSIS	STUDENTS' ANSWERS
INTERACTING IN CLASS	A5TX: "Yes, interacting with the teacher, asking questions." A15TX: "Yes, giving my opinion" A14TY: "Yes, because I was saying some things." A3TZ: "Yes, with participation with questions and answers."
MANIPULATING EXPERIMENTS	A12TX: "Yes, helping to do the practicals and it was very interesting." A12TY: "Yes, I tied a string to my notebook and tried to make the string the same as the other one." A5TZ: "Yes, I helped the teacher when he had to pull the catapult." A6TW: "Yes, helping the teacher, it's very cool, interesting and curious."
PARTICIPATION IN EXAMPLE AND/OR ACTIVITY WORKED ON BY THE RESEARCHER	A3TX: "Yes, I took part in the calculation activity." A1TZ: "Yes, by taking the top sheet off the mobile phone." A7TW: "Yes, by pulling the paper."

Table 17- Categories of analysis A, B, C of the fifth question.
Source: Data collection instrument.

The three categories of analysis and the answers in chart 17 show that they participated and were actively involved in the construction of the lesson, whether by expressing their opinions, solving the activities or collaborating in the execution of the experiments.

According to Carvalho et al. (2007), one of the principles of constructivist teaching related to the role of the teacher in the classroom, which is quite different or even opposed to traditional teaching, consists of interaction.

teacher-student. Thus, in constructivist teaching, the teacher-student relationship is more flexible and collaborative.

Because in this concept of teaching, the student is seen as an individual who "brings with them a 'baggage' of different knowledge, prior knowledge, that is, the structure of knowledge that each one already possesses [...]" (PEREIRA; PADILHA, 2014, p. 7), thus clashing with "traditional teaching" which conceived of the student as being a blank slate, that is, not possessing previous ideas when they arrived at school.

In this context, Tauceda; Nunes and Del Pino (2011, p. 137) criticise the traditional approach to teaching that does not value students' prior knowledge, which "does not produce their knowledge, does not elaborate it, does not rationalise it, they simply passively receive knowledge that is alien to their experiences".

Thus, in the constructivist approach, the student is invited to interact and express their opinion during the course of the lessons, after which it is up to the teacher to re-signify the students' previous knowledge, i.e. that knowledge assimilated from their experiences that a priori does not match the scientific literature, and then subsequently takes on a new meaning, now of a scientific nature.

CATEGORY OF ANALYSIS	STUDENTS' ANSWERS
JUST ATTENDED THE CLASS (PAYING ATTENTION)	A10TX: "No, I just observed the experiment that was done." A10TY: "Yes, I helped by paying attention in class and not getting in the way." A8TZ: "I didn't take part, but I watched and just by watching I learnt and understood a lot." A18TW: "No, but I paid attention, I saw my colleagues doing the practical activities and I learnt."
	A7TX: "No, because I didn't want to, but I have a lot of fun in those classes."
NON-PARTICIPATION	A14TW: "no, but I've had the chance because the teacher gives everyone the opportunity" A20TW: "(...) I didn't want to just observe."
NON-PARTICIPATION DUE TO LACK OF OPPORTUNITY	A11TY: "No, I haven't taken part once! I wanted to take part." A17TY: "No, because they didn't ask me to take part." A2TZ: "no, because other students were asked to do it." A13TW: "no, because there were too many people to debate"

Chart 18 - Categories of analysis D, E, F of the fifth question. **Source:** Data collection instrument.

In the light of the precepts of Araújo and Abib (2003, p. 190) "It is believed that, in general, the appropriate use of different experimental methodologies [...] can enable the formation of an environment favourable to the learning of various scientific concepts [...]".

In fact, according to the data shown in table 18, the students didn't actively "participate" in the course of the lesson for personal reasons or because they didn't have the opportunity due to the length of the lesson. This leads us to infer that the theoretical-practical lessons provided an environment in which the students became more involved with the lesson and did not show any aversion to the activities.

Coelho and Nunes (2003, p. 39) mention that "experimental activities are enriching for the student, since they give real meaning to the abstract and formal world of languages". Therefore, it is necessary to reflect on the fact that experimental activities arouse the student's curiosity in class, while the classroom environment changes in the student's eyes, leading them to pay attention to the phenomena being worked on and their assimilation through a more concentrated gaze.

In the sixth question, **55** answers were analysed from all the students in the four classes. They were grouped according to the categories below expressed in the graph in relation to the percentage of student responses.

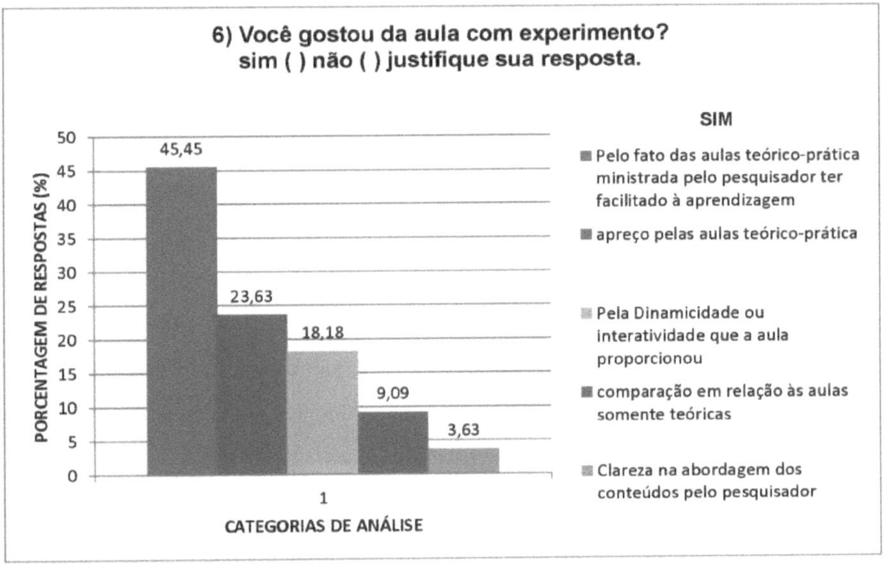

Figura 9 - Graph of the sixth question in the questionnaire applied to students. **Source**: Data collection instrument.

Of the total number of responses analysed, **10** were grouped in the category "the dynamism or interactivity that the class provided", 25 in the category "the fact that the theoretical-practical classes given by the researcher facilitated learning", 5 in the category "comparison with theoretical classes", 13 in the category "appreciation for theoretical-practical classes" and **2** grouped in the category "clarity in the approach to the content".

CATEGORY OF ANALYSIS	STUDENTS' ANSWERS
THE DYNAMISM OR INTERACTIVITY THAT THE LESSON PROVIDED	A2TX: "Yes, because the class interacted with the lesson." A3TY: "Yes, it was very nice, some of my colleagues took part and I think they liked it." A4TZ: "Yes, I was able to learn by playing." A8TW: "yes, because we we relate to each other more (...)"

Chart 19 - Category of analysis A of the sixth question. **Source**: Data collection instrument.

Araújo and Abib (2003) point out, based on a review of various works published in national journals in the area involving experimentation in the teaching and learning process in Physics, between 1992 and 2001, two fundamental aspects for which they believe in the efficiency of this didactic strategy, one of which we highlight here, that experimental activities, "[...] have the capacity to stimulate the active participation of students, arousing their curiosity and interest, favouring an effective involvement with their learning" (ARAÚJO; ABIB, 2003, p. 190-191).

In fact, the data shown in table 19 shows that the students said they liked the theoretical-practical lesson because they took part in the activities and the fact that it made the lesson more dynamic. However, it was clear that these premises added significantly to the teaching and learning process in Physics, and it is clear that this could be an alternative for teachers to combat the challenges of teaching this subject in schools.

CATEGORY OF ANALYSIS	STUDENTS' ANSWERS
BECAUSE THE THEORETICAL-PRACTICAL CLASSES TAUGHT BY THE RESEARCHER FACILITATED	A11TX: "Yes, because it helped me understand better." A18TY: "Yes, because it was nice, I was able to fully understand and observe carefully."

LEARNING.	A2TZ: "Yes, because I understood it better." A3TW: "yes, I can understand Newton's laws better"

Table 20 - Analysis category B of the sixth question.
Source: Data collection instrument.

According to Borges (2002, p. 294) "Science teachers, both in primary and secondary schools, generally believe that improving teaching involves introducing practical lessons into the curriculum". Following this principle, it is pertinent to infer, according to the data shown in table 20, that this strategy associated with the theoretical lesson favoured the students' learning.

It is therefore believed that because physics is an experimental science, practical activities associated with theoretical lessons need to be implemented and valued in schools, with the aim of being an additional pedagogical alternative, in order to add to the construction of differentiated science teaching that provides better conditions for minimising the difficulties that students encounter with this subject.

CATEGORY OF ANALYSIS	STUDENTS' ANSWERS
COMPARISON WITH THEORETICAL CLASSES ONLY	A6TY: "Yes, it's better to do than just watch." A12TY: "Yes, because it was very nice, it was different from the other lessons." A3TZ: "Yes, because it's much better than just explaining without anything." A20TW: "yes, it's a different class from the others"

Chart 21 - Analysis category C of the sixth question.
Source: Data collection instrument

In this category, shown in table 21, it was noted that the students liked the theoretical-practical lesson because it was a different methodology from what they were used to seeing. Thus, it can be seen that this didactic strategy, when well designed and planned, helps to establish a favourable environment for those who think and teach science in a different way.

CATEGORY OF ANALYSIS	STUDENTS' ANSWERS
APPRECIATION FOR THEORETICAL	A11TY: "Yes, I liked it very much, because he made a catapult and one with a trolley and it was very interesting." A20TY: "yes, it was nice, interesting"

AND PRACTICAL LESSONS	A7TZ: "yes, because it was very good." A19TW: "yes, because it was really cool to see physics in practice"

Table 22 - Analysis category D of the sixth question.
Source: Data collection instrument.

It was noted that the students said that the experiments made the lesson more attractive and interesting, as shown in table 22. In view of this, Souza et al. (2012, p. 4) mention that "[...] if the activities developed in the classroom do not encourage interaction between the individual and the object of study, the student may not find a reason to do them".

And in this sense, the authors also mention that experimental activities can be another didactic tool to stimulate and instigate students. Therefore, in an attempt to bring students closer to physical phenomena and laws with the help of experiments, it was found that these practices proved effective in getting students to show greater involvement and interest in the lessons.

CATEGORY OF ANALYSIS	STUDENTS' ANSWERS
CLARITY IN THE RESEARCHER'S APPROACH TO THE CONTENT	A15TY: "Yes, because it was well worked out and explained." A19TY: "Yes, it was well explained with the experiments, I liked it a lot."

Chart 23 - Category of analysis E of the sixth question. **Source**: Data collection instrument.

In this category, shown in table 23, we can see that the clarity factor in the presentation of the content influenced the way the students conceived the lesson. Therefore, when the teacher has this kind of attitude in the classroom, as the data shows, it can add significantly to bringing students closer to the subject.In the seventh question, **65** answers were analysed from the total number of students in the four classes. They were grouped according to the categories below, expressed in the graph in relation to the percentage of student responses.

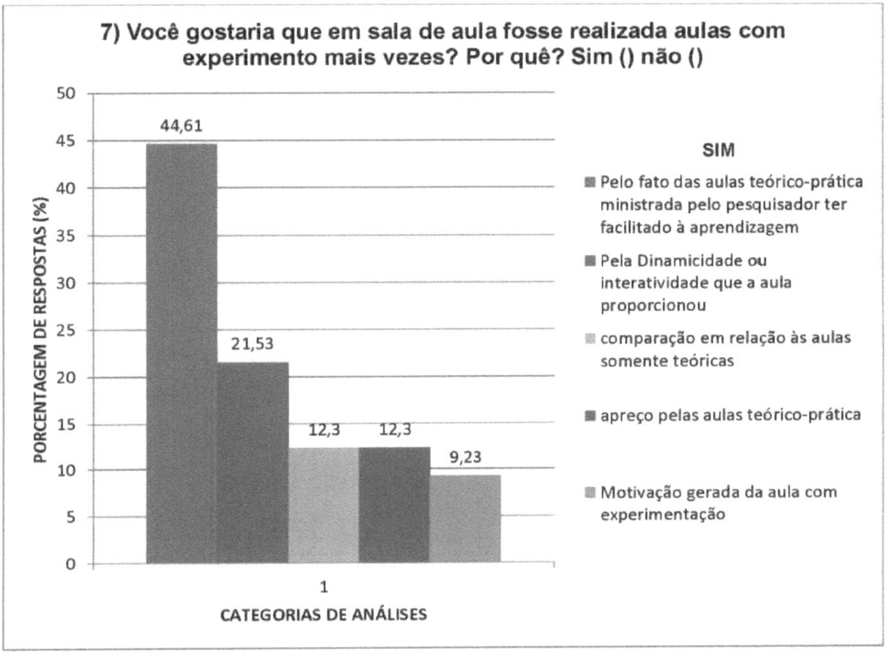

Figura 10 Graph of the seventh question of the questionnaire applied to the students. SOURCE: Data collection instrument.

Of all the answers analysed, 14 were grouped in the category "the dynamism or interactivity that ***the*** class provided", 06 in the category "motivation generated by the class with an experiment", 29 in the category "the fact that the theoretical-practical classes given by the researcher facilitated learning", 08 in the category "appreciation for theoretical-practical classes" and 08 were grouped in the category "comparison with theoretical classes only".

CATEGORY OF ANALYSIS	STUDENTS' ANSWERS
THE DYNAMISM OR INTERACTIVITY THAT THE LESSON PROVIDED	A2TX: "Yes, because the class doesn't just watch the lesson, it interacts." A17TY: "Yes, because it makes lessons more fun (...)" A4TZ: "Yes, to make the lesson more fun." A10TW: "yes, because the students interact more, participate more."

Chart 24 - Analysis category A of the seventh question. **Source**: Data collection instrument.

It is worth noting that the fact that the students were asked to manipulate the actions of the experiments supported by the theory covered in the lecture by the researcher significantly aroused the students' interest and involvement in the class, according to chart 24.

Gádea and Dorn (2011, p. 113) argue that "by carrying out 'playful' activities with a constructivist view of science, we are giving children the opportunity to exercise their skills and develop their critical and reflective capacity [...]".

On the other hand, Lima et al. (2009, p. 8) point out that experimentation encourages the "[...] ability to operate co-operatively [...]". In view of this, it should be emphasised that supporting methodologies that involve experimentation and greater student interaction in the delivery of lessons is a viable alternative for awakening students' interest and taste for science.

CATEGORY OF ANALYSIS	STUDENTS' ANSWERS
MOTIVATION GENERATED BY THE LESSON WITH EXPERIMENTATION	A3TX: "Yes, because I want to learn more." A3TZ: "Yes, it could improve students' attention." A5TZ: "yes, because the student becomes much more engaged with the experiences." A8TW: "yes, because it encourages more learning"

Chart 25 - Category of analysis B of the seventh question. **Source**: Data collection instrument.

As shown in table 25, we observed that after the theoretical-practical lessons, some of the students felt enthusiastic and showed greater motivation towards the subject in question. In this sense, Galiazzi et al. (2001, p. 254) state that one of the factors usually listed by teachers and students when referring to the effects of experimental activities "[...] is its motivating character [...]".

In the view of Lima et al. (2009, p. 8) "[...] the playful, enlightening and challenging nature of experimentation favours the development of student self-esteem and appreciation for the educational process in which they are involved [...]". In this context, when experimentation is used well in the teaching and learning process, it aims to change the concept of "transmitter" teaching to give way to the establishment of differentiated working attitudes with a view to encouraging students to learn scientific concepts.

CATEGORY OF ANALYSIS	STUDENTS' ANSWERS
BECAUSE THE THEORETICAL-PRACTICAL CLASSES TAUGHT BY THE RESEARCHER FACILITATED LEARNING.	A10TX: "Yes, because the more experiments we carry out, the better our knowledge will be." A18TY: "Yes, because it's much better to understand, and with the explanations we have a clearer concept of the subject." A7TZ: "Yes, because it's easier to understand with an experiment." A11TW: "Yes, because it's easy to understand the subject."

Chart 26- Category of analysis C of the seventh question.
Source: Data collection instrument.

When Pozo and Gómez Crespo (2009, p. 192) talk about the difficulties that students encounter in learning physics concepts, it is fundamentally due to "(...) the differences and apparent contradictions between the idealised world that science presents and the real world that the student observes", the authors also stress that "(...) understanding the physics taught at school would require overcoming the restrictions imposed by the students' own theories" (POZO; CRESPO, 2009, p. 194).

These authors believe that when the teacher provides a link between the knowledge that students bring to school, and that which they observe and elaborate on a day-to-day basis, and the scientific concepts of physics, they will be more successful in teaching the subject.

Because, still inspired by the authors above, when students arrive at school they relate what they previously knew to the new concepts worked on by the teacher and, in most cases, they enter into cognitive conflict when trying to materialise the new knowledge that is being shown to those they had before arriving at school.

Bearing in mind that many of the phenomena that occur in the student's daily life are explained by physical principles which require cognitive maturity and the ability to imagine in order to understand them.

On the other hand, Arruda and Laburú (2009, p. 64) emphasise that in physics classes "(...) the function of the experiment is to adapt theory to reality". Based on this assumption, and the fact that the curricular component in question is "complex" for students to learn, the data shows that experimentation can indeed help to clarify some physical principles for students. While handling the experiment, the concepts become more "materialised" in the student's cognitive field, and with the teacher's help, the students associate the theories with the actions of the practical activity and this makes their learning more satisfactory. As a result, it influences the way students see and learn the subject.

CATEGORY OF ANALYSIS	STUDENTS' ANSWERS
APPRECIATION FOR THEORETICAL AND PRACTICAL LESSONS	A12TX: "Yes, I find the lessons more interesting." A3TY: "(...) and it would make the lessons more interesting." A18TW: "Yes, it's a nice, interesting thing, I hope it's repeated many, many times" A19TW: "yes, because it's really cool"

Chart 27 - Analysis category D of the seventh question.
Source: Data collection instrument.

Buch and Schroeder (2013, p. 74) point out that "learning science in a meaningful way presupposes the effective participation of students in their knowledge-building processes". Imbemon et al. (2009, p. 84) argue that "[...] the use of experimental activities during lessons allows students to acquire the curiosity necessary for learning science [...]".

In fact, it is clear that the teacher's efforts to work with methodologies using experimentation have been reflected in the way the content is explored, overcoming the "abstraction" that is required of the student when faced with only theoretical presentations, in order to provide a different environment that makes the lessons more attractive to the student, as the students' responses in table 27 elucidate.

CATEGORY OF ANALYSIS	STUDENTS' ANSWERS
COMPARISON WITH THEORETICAL CLASSES ONLY	A11TY: "Yes, because the students would pay more attention and take part in the lessons too. I want to!" A15TY: "Yes, because writing a lot is boring." A19TY: "Yes, because it would not only help us to remember at the time of the exam, but also to seek knowledge." A1TZ: "Yes, because we don't understand much just by explaining without practical lessons."

Chart 28 - Category of analysis E of the seventh question. **Source**: Data collection instrument

Imbernon et al. (2009, p. 81-82) point out that one of the biggest problems that teachers tend to overcome in relation to science teaching is often due to "[...] the lack of meaning that scientific content presents to the student and the absence of a logical connection between theoretically developed content and the real world in which the student lives".

According to the author, "[...] if the theoretical content presented is not contextualised in their day-to-day lives, the student, even if they 'understand' the concepts, will not know how to apply them properly, making learning mechanical and meaningless [...]" (IMBRENON et al., 2009, p. 84).

In this sense, we advocate a different kind of science teacher who provides students with content that is interconnected with the outside world. According to the students' responses, experimentation would be an attempt to stimulate their taste for science, with the aim of overcoming the intense exposition of content without "meaning", which is often the case in schools.

In the eighth question, **31** answers were analysed from all the students in the four classes. They were grouped according to the categories below expressed in the graph in relation to the percentage of students' answers.

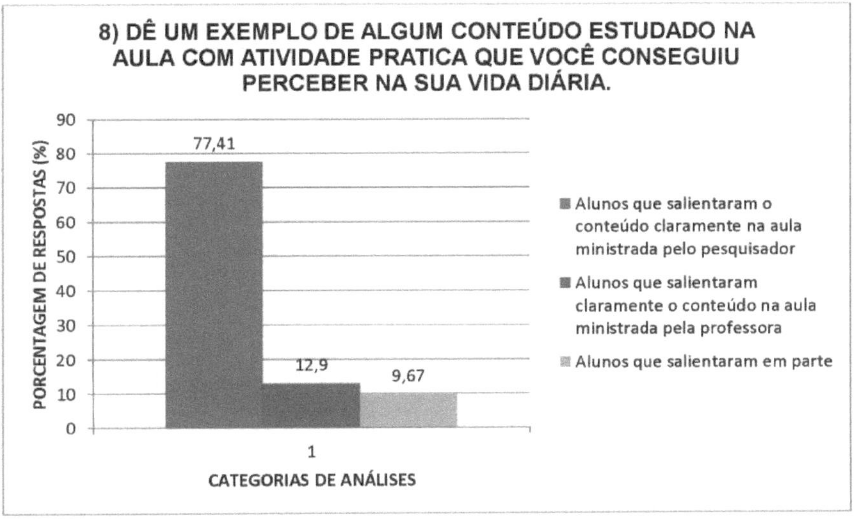

Figura 11 Graph for the eighth question.
Source: Data collection instrument.

Of the total responses analysed, **24** were grouped into the category "students who clearly emphasised the content in the lesson taught by the researcher", 04 into the category "students who clearly emphasised the content in the lesson taught by the teacher" and 03 were grouped into the category "students who emphasised it in part".

CATEGORY OF ANALYSIS	STUDENTS' ANSWERS
STUDENTS WHO CLEARLY EMPHASISED THE CONTENT IN THE LESSON TAUGHT BY THE RESEARCHER.	A14TX: "from the pram that if you don't wear a seatbelt you can be thrown out of the car." A9TY: "Car accidents, people are thrown in front of the car." A4TZ: "Traffic accidents, how and why people fly." A17TW: "Newton's laws that every action has a reaction"
STUDENTS WHO EMPHASISED THE CONTENT CLEARLY, IN THE CLASS TAUGHT BY THE TEACHER	A12TY: "physical quantities." A17TY: "physical quantities." A5TZ: "about electricity" A4TW: "the time that is at all times in our lives"
STUDENTS WHO PARTLY EMPHASISED	A2TY: "it was the explanation of physics itself where they explain the phenomena of nature" A5TY: "Yes, but I can't remember the name." A6TY: "(...) I understand, but I can't remember the name (...)"

Table 29- Categories of analyses A, B, C of the eighth question.
Source: Data collection instrument.

It was noteworthy that in this question 90, 3 per cent of the answers pointed to some content that the students perceive in everyday life. We should also point out that, based on the theoretical-practical classes, 77, 41% of the answers analysed showed that they had managed to relate the content studied to everyday reality.

Thus, it was observed that the "bump" experiment, in which Newton's first law was worked on and analogies were made with everyday situations involving traffic accidents, was relevant for the content to gain meaning for the students, according to the answers A14TX; A9TY and A4TZ in table 29 - category of analysis A.

On the other hand, compared to the second question, the rate of students who did not clearly highlight some of the content studied was 9.67 % (see figure - 11) compared to 53.65 % (see figure - 5) of the answers analysed for the second question.

In this way, it is clear that when content gains meaning for students, they feel more enthusiastic about learning when they can see that many of the phenomena that occur in everyday life can be explained by physical laws, "[...] given that Physics is part of our daily lives and even if we don't like it, we are part of a 'physical world'" (MOREIRA; PENIDO, 2009, p. 4).

It is therefore important that science content is worked on close to the students' reality in order to stimulate their curiosity and make them realise that science and technology influence and shape society.

And society influences and shapes science and technology in order to contribute to the formation of a critical positioning and development of students, which we believe that experimental activities combined with other didactic-methodological skills of the teacher can be an alternative to try to materialise this principle.

CHAPTER 6

FINAL CONSIDERATIONS

This research showed that physics lessons in the 9th year of primary school are still predominantly based on theoretical lessons, with the textbook as the main resource.

We found that in this context, students are conceived as individuals who don't have spontaneous knowledge, while this type of methodology values the direct exposition of ready-made and consolidated concepts on the blackboard, which students must memorise and solve the exercises. On the other hand, we note the teacher's attempt to provide a different environment in class, such as taking the students to watch videos about the subject and holding a debate to discuss the content.

However, if the teacher does not provide an environment where students transform their previous conceptions into scientific ones (AUSUBEL, 2003), the attempt to provide a different environment can turn into dull lessons in which students are not encouraged to give an opinion or relate the concepts being worked on to their reality.

In this sense, it is clear that experimentation used in the teacher's methodology, taking into account the reality and level of development of the students, can be a viable alternative for alleviating the difficulties that many students encounter in learning the subject of Physics.

It is clear that there is no ready-made recipe or teaching method that can solve all the problems arising from science teaching, but it was observed that practical activities stimulate student participation and make lessons more attractive, reflecting the teacher's attempt to plan activities and provide experiments that can change the classroom scenario.

The factor "clarity in the approach to the content" was emphasised by many students as a positive premise for studying physics. This shows us the challenges arising from the teacher's practice, because when the teacher tries to develop a differentiated posture in his classes, suggesting the articulation of theory and practice in the teaching and learning process, his work will be doubled.

We are therefore left with the reflection that despite being a topic that is widely disseminated by researchers and also receiving massive criticism about its conceptions and ways of being used in the classroom, experimentation is still not widely used in schools.

It is therefore believed that this type of methodology favours student learning when combined with the teacher's other didactic-methodological skills, leading to the establishment of a classroom scenario where scientific concepts are worked on in a less "crystallised" and mechanical way.

REFERENCES

ARAÚJO, M. S. T; ABIB, M. L. V. S. As atividades experimentais no ensino de Física: Diferentes enfosques, diferentes Finalidades. **Revista Brasileira de Ensino de Física**. v. 25, n, 2. 2003.

AGOSTINI, V. W; DELIZOICOV, N. C. Didactic experimentation in teaching Fundamental: Impasses and challenges. In: ENCONTRO NACIONAL DE PESQUISA EM EDUCAÇÃO EM CIÊNCIAS - ENPEC, 7, 2009, Florianópolis. **Electronic Annals**... Available at: <http://www.google.com.br/url?sa=t&rct=j&q=&esrc=s&source=web&cd=1&ved=0CB0QFjAA&url=http%3A%2F%2Fposgrad.fae.ufmg.br%2Fposgrad%2Fviienpec%2Fpdfs%2F1225.pdf&ei=sJZnVKbIGciZgwTtvYToDQ&usg=AFQjCNEKY0Uz09QaCUcl U9qdE6Va5zy_Q&sig2=eIBRLpnDLQOOXyhPJsvvrw&bvm=bv.79142246,d.e XY>. Accessed on: 22 October 2014.

ARRUDA, S. M; LABURÚ, C. E. Considerations on the role of experiments in science teaching. In: NARDI, R. (Org.) **Questões atuais no ensino de Ciências**.2 ed. São Paulo: Escrituras, 2009. p. 59 - 66.

ATAIDE, M. C. E. S; SILVA, B. V. C. Teaching Methodologies in Science: Contributions of Experimentation and the History and Philosophy of Science. **Holos**, v. 4, p. 171 - 181, 2011.

AUSUBEL, D. P. **Acquisition and retention of knowledge**: a cognitive perspective. Lisbon: Plátano, 2003. 219 p.

BARDIN, L. **Content Analysis**. São Paulo: Edições 70, 2011. 279 p.

BARBIER, R. **Action Research**. Brasília: Liber Livro Editora, 2007. 157 p.

BORGES, A. T. New directions for the school science laboratory. **Cad. Bras. Ens. Fis.**, v. 19, n. 3, p. 291 - 313. 2002.

BOURSCHEIDT, L. Action research: an alternative for music education research in Brazil. In: ANPPOM CONGRESS, 17, 2007, São Paulo. **Electronic Proceedings**... São Paulo: Editora da Universidade Estadual de São Paulo, 2007. Available at: <www.anppom.com.br/.../anaiscongresso_anppom.../educacao_musica... >. Accessed on: 20 June, 2014.

BRAZIL, Secretariat of Basic Education. **National Curriculum Parameters: Natural Sciences**. Brasília: MEC/SEF, 1998.

BUCH, G. M; SCHROEDER, E. Clubes de Ciências e Alfabetização Científica: Concepções dos professores coordenadores da rede municipal de ensino de Blumenau (SC). **Experiências em ensino de Ciências**. v. 8, n. 1, p. 72-86. 2013.

CARVALHO, A. M. P; VANNUCCHI, A. I; BARROS, M. A; GONÇALVEZ, M. E. R; REY, R. C. **Ciências no Ensino Fundamental**: O Conhecimento Físico. São Paulo: Scipione, 2007.

CARVALHO, A. M. P. Structuring criteria for the teaching of Science. In: CARVALHO, Ana Maria Pessoa de (Org.). Ensino de Ciências: **unindo a Pesquisa e Prática**. São Paulo: Pioneira Thomson Learning, 2004. p. 1-17.

COELHO, S. M; NUNES, A. D. O papel da Experimentação no ensino da Física. **Cad. Bras. Ens. Fís.**, v. 20, n. 1, p. 30-42, 2003.

DALFOVO, M. S; LANA, R. A; SILVEIRA, A. Métodos Quantitativos e Qualitativos: Um Resgate Teórico. **Revista Interdisciplinar Cientifica Aplicada**, Blumenau, v. 2, n. 4, p. 01-13, 2008.

DELIZOICOV, D; ANGOTTI, J. A; PERNAMBUCO, M. M. **Ensino de Ciências**: Fundamentos e Métodos. 2 ed. São Paulo: Cortez, 2007, p. 12.

DUTRA, L. V. Centro Universitário das Faculdades Associadas de Ensino - Unifae: **Elaborando Projeto de Pesquisa**. São Paulo, 2006.

FRANCO, M. A. S. Pedagogy of action research. **Educação e Pesquisa**, São Paulo, v. 31, n. 3, p. 483-502, Sep./Dec. 2005.

FURMAN, M. **O ensino de Ciências no Ensino** Fundamental: Colocando as Padras Fundamentais do Pensamento Cientifico. Brazil, São Paulo: Sangari, p. 01-18, 2009.

GADÉA, S. J. S; DORN, R. C. Scientific literacy: thinking about science learning in the early grades through experimental activities. **Experiências em Ensino de Ciências** (UFRGS), v. 6, p. 113131, 2011.

GALIAZZI, M. C; ROCHA, J. M. B; SCHMITZ, L. C; SOUZA, M. L; GIESTA, S; GONÇALVES, F. P. Objetivos das atividades experimentais no Ensino Médio: A pesquisa coletiva como modo de formação de professores de Ciências. **Ciência e Educação**. v. 7, n.2, p. 249-263, 2001.

GASPAR, A. Compreendendo a Física: Ensino Médio - **Newton's laws**. 1 ed. São Paulo: Ática, p. 116-129, 2010.

GERHARDT, T. E; RAMOS, I. C. A; RIQUINHO, D. L; SANTOS, D. L. Structure of the Research Project. In: GERHARDT, T. E; SILVEIRA, D. T. (Orgs.) **Métodos de Pesquisa**. Porto Alegre: UFRGS publishing house, 2009.

IMBERNON, R. A. L; TOLEDO, M. C. M; HONÓRIO, K. M; TUFAILE, A. P. B; VARGAS, R. R. S; CAMPANA, P. T; FALCONI, S; INFANTE- MALACHIAS, M. E. Experimentation and Interactivity (HANDS-ON) in Science Teaching: Practice in Pedagogical Praxis. **Experiências em Ensino de Ciências**. v. 4, n. 1, p. 79-89. 2009.

JESUS, V. L. B.; MACEDO JUNIOR, M. A. V. A Discussion on Hydrodynamics Using PET Bottles. **Revista Brasileira de Ensino de Física**. v. 33, n. 1, 1507, 2011.

LABURÚ, C. E.; SILVA, O. H. M.; SALES, D. R. Conceptual Overcoming of High School Students in Measurement from the Questioning of a Problematic Experimental Situation. **Revista Brasileira de Ensino de Física**, v. 32, n. 1, p. 1402, 2010.

LIMA, V. M. R; HARRES, J. B. S; BORGES, R. M. R; ROCHA FILHO, J. B. Presentation and Evaluation of Material for Support and Experimentation in Physics Teaching. **Experiências em Ensino de Ciências**, v. 4, n. 1, p. 7-22, 2009.

MARTINS, A. F. P. Supervised Internship in Physics: The Pulse Still Pulses. **Revista Brasileira de Ensino de Física**, v. 31, n. 3, p. 3402, 2009.

MEDEIROS, L. G. S. **Why Learn Natural Sciences?** In: MEDEIROS, L. G. S. (Org.) Metodologia e Instrumentação para o Ensino de Ciências Naturais. Cadernos Cb Virtual 5. João Pessoa: Ed. Universitária, p. 165- 166. 2010.

MEDEIROS, A. A; LEITE, R. D. O; OLIVEIRA, N. S. M. Physics for diversity: from concepts to making accessible material. In: IV MULTIDISCIPLINARY BRAZILIAN

CONGRESS ON SPECIAL EDUCATION, 2007, Londrina. **Electronic annals**... Available at: <http://www.uel.br/eventos/congressomultidisciplinar/pages/arquivos/anais/200 7/297.pdf>. Accessed on: 22 October 2014.

MEGID NETO, J.; PACHECO, D. Pesquisa Sobre o Ensino de Física no Nível Médio no Brasil: Concepção e Tratamento de Problemas em Teses e Dissertações. In: NARDI, R. (Org.) **Pesquisas em Ensino de Física**. 2 ed. São Paulo: Escrituras, 2001. p. 17.

MINAYO, M. C. S. O Desafio do Conhecimento - **Pesquisa Qualitativa em Saúde**. São Paulo: Hucitec, 2008.

MOREIRA, A. C. S; PENIDO, M. C. M. On proposals for the use of experimental activities in the teaching of Physics. In: ENCONTRO NACIONAL DE PESQUISA EM EDUCAÇÃO EM CIÊNCIAS - ENPEC, 7, 2009, Florianópolis. **Electronic Proceedings**... Available at: <http://www.google.com.br/url?sa=t&rct=j&q=&esrc=s&source=web&cd=1&ved=0CB0QFjAA&url=http%3A%2F%2Fposgrad.fae.ufmg.br%2Fposgrad%2Fviienpec%2Fpdfs%2F814.pdf&ei=L5dnVIiMMsKpgwSJ4YGIBg&usg=AFQjCNGWjVh3HbeYz8RpFhno_Wfyc5x4g&sig2=Q6hJpUi5qzcx1oLBimfRfA&bvm=bv.79142246,d.eXY>. Accessed on: 22 October 2014.

NARDI, R. **Pesquisas em Ensino de Física**. 2 ed. São Paulo: Escrituras, 2001.

OLIVEIRA, M. M. L; COSTA, R. C; SOLETO, D. G; ROCHA FILHO, J. B. Physics Experimental Practices in the Context of Teaching Through Research: A Reflection. **Experiences in Science Teaching**. v. 5, n. 3, 29-38. 2010.

PACHECO, D. Experimentation in science teaching. **Science and Teaching**. 1997.

PORTO, A.; RAMOS, L.; GOULART, S. **Um olhar Comprometido Com o Ensino de Ciências**. Belo Horizonte: FAPI, 2009.

POZO, J. I; GÓMEZ CRESPO, M. A. **Learning and teaching science: from everyday knowledge to scientific knowledge**. 5 ed. Porto Alegre: Artmed, 2009.

PEREIRA, B. B. **EXPERIMENTATION IN SCIENCE TEACHING AND THE ROLE OF THE TEACHER IN THE CONSTRUCTION OF KNOWLEDGE**. Cadernos da FUCAMP, v. 9, p. 83-92, 2010.

PEREIRA, G. F. S; PADILHA, E. C. P. The importance of prior knowledge in science teaching - In the light of David Ausubel's theory of meaningful learning. In: **Anais...** V NATIONAL MEETING ON SIGNIFICANT LEARNING - ENAS V, 2014, Belém- Pa.

PEREIRA, G. F. S; RIBEIRO, E. O. R; FREITAS, N. M. S. **Appropriation of scientific knowledge**: an approach to transgenic foods. Belém: Novas Edições Acadêmicas, 2014. 96 p.

RODRIGUES, M. A.; TEIXEIRA, F. M. O Ensino de Física nas Séries Iniciais do Ensino Fundamental na Rede Municipal de Ensino do Recife Segundo Os Seus Docentes. **Revista Brasileira de Ensino de Física**. v. 33, n. 4, p. 4401, 2011.

SANTOS, A. R. Metodologia Cientifica: A **Construção do Conhecimento**. 6 ed. rev. Rio de Janeiro: DP&A, 2006.

SANTOS, J. C; GOMES, A. A; PRAXEDES, A. P. P. Teaching Physics: from Teaching Methodology to Learning Conditions. In: **Electronic Annals**... V ENCONTRO DE PESQUISA EM EDUCAÇÃO EM ALAGOAS, 2010, Maceió. Available at: <http://dmd2.webfactional.com/media/anais/ENSINO- DA-FISICA.pdf>. Accessed on: 22 October 2014.

SILVA, M. O. L. The teaching of natural sciences and teaching practice. In: **Anais eletrônicos**... VII ENCONTRO NACIONAL DE PESQUISA EM EDUCAÇÃO EM CIÊNCIAS - ENPEC, 7, 2009, Florianópolis. Available at: <www.ufpi.br/...1/13_Maria%20Oenide%20Lima%20da%20Silva.pdf>. Accessed on: 22 October 2014.

SILVA, M. N. M; ROCHA FILHO, J. B. The current role of experimentation in physics teaching. In: **Electronic Proceedings**... XI SALÃO DE INICIAÇÃO CIENTÍFICA - PUCRS, 2010. Available at: <http://www.google.com.br/url?sa=t&rct=j&q=&esrc=s&source=web&cd=1&ved=0CB0QFjAA&url=http%3A%2F%2Fwww.pucrs.br%2Fedipucrs%2FXISalaoIC%2FCiencias_Exatas_e_da_Terra%2FFisica%2F84372MAURICIONOGUEIRAMACIELDASILVA.pdf&ei=cplnVMablYuhgwT_3YKgCA&usg=AFQjCNHlTZvh0RnUN9QXeOc8vhtC8OIEZQ&sig2=_XoxiakuMBAJCtpMHMSf-g&bvm=bv.79142246,d.eXY>. Accessed on: 22 October 2014.

SOUZA, N.F.; CAMPOS, B.S.; FERNANDES, S.A.; RAGNI, A.C.P.B. Física Para Crianças: Abordando Conceitos Físicos a partir de Situações-Problema. **Revista Brasileira de Ensino de Física**. v. 34, n. 1, 1402. 2012.

SOUZA, T. C. F; HEINECK, R. Researching the Different Learning Assessment Methods and the Use of Their Didactic Resources - From the Perspective of Physics Educators. **Experiências em Ensino de Ciências**. v. 1, n. 2, 01 - 09. 2006.

TAUCEDA, K. C; NUNES, V. M; DEL PINO, J.C. A Epistemology/Methodology of the Student Researcher in Science Education. **Experiences in Science Teaching**. v. 6, n. 3, 133 - 141, 2011.

ZÔMPERO, A. F; PASSOS, A. Q; CARVALHO, L. M. A Docência e as Atividades de Experimentação no Ensino de Ciências nas Séries Iniciais do Ensino Fundamental. **Experiences in Science Teaching**, v. 7, n. 1, p. 43-54, 2012.

APPENDICES

APPENDIX 1

Questions worked on with the students in the theoretical-practical class.

1)A 15 kg body was pushed horizontally in a rightward direction and had an acceleration of 3 m/s2. What was the FORCE applied to the body?

2) A force of 10 N was applied horizontally in a leftward direction to a 5 kg body. What acceleration did this body undergo?

3) A wooden box is accelerated by 5 m/s2 and the force applied to it is 15 N. What is the MASS of this body?

APPENDIX 2

Date: ___/___/___

Structured observation guide for theoretical and theoretical-practical classes.

1) Capacity for initiative and student participation in lessons.
2) Students' questions about the content.
3) Curiosity about the content.
4) Difficulties presented in relation to the lessons.
5) Students' esteem towards the lessons.

APPENDIX - 3

Questionnaire administered to 9th grade students.

Class: ______

Student: ____ **Date:**

___/___/___

1) Do you like the way the teacher works with the subjects in physics lessons? Why do you like it?

Yes () No ()

2) Can you see the physics content that is taught in the classroom in your daily life? Justify your answer.

Yes () No ()

3) Has the teacher ever carried out any kind of experiment (practical activity) in the classroom?If there was an experimental lesson, did you enjoy it?Justify your answer.

Yes () No ()

IN RELATION TO THE PHYSICS LESSONS YOU WORKED ON WITH PRACTICAL ACTIVITIES, ANSWER THE FOLLOWING QUESTIONS:

4)Do you understand physics content better when the teacher just explains it or when they explain it and carry out experiments (practical activities) in the classroom? Why?

5) Have you ever taken part in an experiment (practical activity) in class? In what way?

Yes () No ()

6) Did you like the lesson with the experiment (practical activity)? Justify your answer.

Yes () No ()

7) Would you like to have classes with experiments (practical activities) more often?

Why?

Yes ( )   No ( )

8) Give an example of some of the content studied in class with a practical activity that you have realised in your daily life.

ANNEXES

ANNEX 1

INFORMED CONSENT FORM

You are being invited to take part, as a volunteer, in a research project on **"THE RELEVANCE OF EXPERIMENTAL CHARACTERS IN PHYSICS TEACHING FOR THE NINTH YEAR OF EDUCATIONAL EDUCATION".**

Once you have been informed of the following information, if you agree to take part in the study, please sign at the end of this document, which is in two copies. One copy is yours and the other belongs to the researcher responsible. If you refuse, you will not be penalised in any way.

INFORMATION ABOUT THE RESEARCH:

Project title: **"THE RELEVANCE OF THE EXPERIMENTAL CHARACTER IN PHYSICS TEACHING FOR THE NINTH YEAR OF EDUCATIONAL EDUCATION".**

Work supervisor: Specialist Professor Rosivaldo Carvalho Gama Júnior Contact telephone number (including collect calls): (96) 91373482.

The aim of the work is to develop the student's capacity for scientific production, in order to train good professionals in the field of education. The research will be carried out through direct observation, the application of questionnaires to 9th grade students with semi-structured questions and the application of a lesson with experimentation in the classes selected for the research, in this case, if the research is action research according to the data collection.

The research will be carried out between August and October 2014 and does not involve any risk to life and/or morals, nor will it cause any damage, discomfort and/or injury. The young researcher, Josefer Ranchel da Silva Neves, undertakes not to divulge information given by interviewees who withdraw consent to the interview throughout the course of the research; that in this case, the interviewee may withdraw consent at any time during the aforementioned period, and when the information is kept, it will be under the total confidentiality of the interviewee, which in this case, a pseudonym and/or abbreviation of the interviewee's name will be used with indication of the place of the interview.

____

Josefer Ranchel da Silva Neves (Researcher/student)

CONSENT TO THE PERSON'S PARTICIPATION AS SUBJECT

Me, ____ , ID / CPF/ no. , below signed, I agree to take part in the study concerning: **The Relevance of the Experimental Character in Physics Teaching for the 9th Year of Elementary School**, as a subject. I have been duly informed and clarified by the student/researcher **Josefer Ranchel da Silva Neves** about the research, the procedures involved, as well as the possible risks and benefits arising from my participation. I was assured that I could withdraw my consent at any time, without this leading to any penalty or interruption of my monitoring/assistance/treatment.

Place and date: **Macapá/AP, ___/_____/ 2014**

Name of subject or guardian:

____

Signature:

We witnessed the request for consent, clarifications about the research and the subject's acceptance to take part in the research mentioned here.

Witnesses (not connected to the research team):

Name:

Signature:____

Name: ____

Signature:

____

Additional comments:

Kolibri 360 GmbH
Ferdinand-Jühlke-Straße 7
99095 Erfurt, Deutschland
produktsicherheit@kolibri360.de

Druck:
CPI Druckdienstleistungen GmbH
im Auftrag der
Kolibri 360 GmbH
Ferdinand-Jühlke-Str. 7
99095 Erfurt